AF581860

Advances of Spectrometric Techniques in Food Analysis and Authentication

Advances of Spectrometric Techniques in Food Analysis and Authentication

Editor

Daniel Cozzolino

MDPI • Basel • Beijing • Wuhan • Barcelona • Belgrade • Manchester • Tokyo • Cluj • Tianjin

Editor
Daniel Cozzolino
Centre for Nutrition and
Food Sciences, Queensland
Alliance for Agriculture and
Food Innovation,
St. Lucia Campus,
The University of
Queensland,
Brisbane, Australia

Editorial Office
MDPI
St. Alban-Anlage 66
4052 Basel, Switzerland

This is a reprint of articles from the Special Issue published online in the open access journal *Foods* (ISSN 2304-8158) (available at: https://www.mdpi.com/journal/foods/special_issues/Spectrometric_Analysis_Authentication).

For citation purposes, cite each article independently as indicated on the article page online and as indicated below:

LastName, A.A.; LastName, B.B.; LastName, C.C. Article Title. *Journal Name* **Year**, *Volume Number*, Page Range.

ISBN 978-3-0365-6668-9 (Hbk)
ISBN 978-3-0365-6669-6 (PDF)

Contents

Editorial

Advances in Spectrometric Techniques in Food Analysis and Authentication

Daniel Cozzolino

Centre for Nutrition and Food Sciences, Queensland Alliance for Agriculture and Food Innovation, St. Lucia Campus, The University of Queensland, Brisbane, QLD 4072, Australia; d.cozzolino@uq.edu.au

The demand from the food industry and consumers for analytical tools that can assure the quality (e.g., composition) and origin of foods (e.g., authenticity, fraud, provenance) in both the supply and value chains has increased over the past decades. Although there have been advances and improvements in analytical instrumentation and techniques that have excellent diagnostic capabilities, most of the existing routine methods of analysis are considered time-consuming and expensive. These issues have encouraged developments in the application of a wide range of spectrometric techniques, involving, among others, the utilization of vibrational spectroscopy combined with data analytics (e.g., chemometrics).

This Special Issue, "Advances in Spectrometric Techniques in Food Analysis and Authentication", has compiled novel and recent applications of spectrometry-based techniques, including NIR, MIR, NMR, as well as other analytical techniques (e.g., ICP-MS and GC-MS) combined with chemometrics methods, to target issues associated with food analytics and authentication along the food supply and value chains (e.g., fraud, provenance, traceability).

In this Special Issue, Gajek and collaborators [1,2] have shown how the combination of chemometrics with ICP-MS data can be used to authenticate whisky samples. Chavez-Angel and collaborators have also described how vibrational spectroscopy (e.g., Raman and infrared) and thermal analysis can be combined to authenticate extra virgin olive oil [3]. The classification of coffee samples was also reported by combining both NIR and MIR spectroscopy with chemometric methods [4]. The utilization of spectral fingerprints was evaluated as a potential tool to screen the adulteration of traditional and Bourbon barrel-aged maple syrups [5]. The use of portable NIR instrumentation was also reported to discriminate and characterize individual goat muscles [6] and to differentiate meat species using a similarity index [7]. The use of a liquid–liquid microextraction method to enhance the flavor of seafood using GC-MS analysis [8], the analysis of saliva using MIR spectroscopy obtained from a sensory study [9], and the determination of the saponification value of fats and oils using 1H-NMR were also reported [10] in this Special Issue.

Overall, these applications have highlighted the importance of combining rapid analytical methods with chemometrics to improve our knowledge and understanding about foods.

Citation: Cozzolino, D. Advances in Spectrometric Techniques in Food Analysis and Authentication. *Foods* **2023**, *12*, 438. https://doi.org/10.3390/foods12030438

Received: 10 January 2023
Accepted: 14 January 2023
Published: 17 January 2023

Funding: This research received no external funding.

Data Availability Statement: Not applicable.

Conflicts of Interest: The author declares no conflict of interest.

References

1. Gajek, M.; Pawlaczyk, A.; Maćkiewicz, E.; Albińska, J.; Wysocki, P.; Jóźwik, K.; Szynkowska-Jóźwik, M.I. Assessment of the Authenticity of Whisky Samples Based on the Multi-Elemental and Multivariate Analysis. *Foods* **2022**, *11*, 2810. [CrossRef] [PubMed]
2. Gajek, M.; Pawlaczyk, A.; Jóźwik, K.; Szynkowska-Jóźwik, M.I. The Elemental Fingerprints of Different Types of Whisky as Determined by ICP-OES and ICP-MS Techniques in Relation to Their Type, Age, and Origin. *Foods* **2022**, *11*, 1616. [CrossRef] [PubMed]
3. Chavez-Angel, E.; Puertas, B.; Kreuzer, M.; Soliva Fortuny, R.; Ng, R.C.; Castro-Alvarez, A.; Sotomayor Torres, C.M. Spectroscopic and Thermal Characterization of Extra Virgin Olive Oil Adulterated with Edible Oils. *Foods* **2022**, *11*, 1304. [CrossRef] [PubMed]
4. Belchior, V.; Botelho, B.G.; Franca, A.S. Comparison of Spectroscopy-Based Methods and Chemometrics to Confirm Classification of Specialty Coffees. *Foods* **2022**, *11*, 1655. [CrossRef] [PubMed]
5. Zhu, K.; Aykas, D.P.; Rodriguez-Saona, L.E. Pattern Recognition Approach for the Screening of Potential Adulteration of Traditional and Bourbon Barrel-Aged Maple Syrups by Spectral Fingerprinting and Classical Methods. *Foods* **2022**, *11*, 2211. [CrossRef] [PubMed]
6. Hoffman, L.C.; Ingle, P.; Khole, A.H.; Zhang, S.; Yang, Z.; Beya, M.; Bureš, D.; Cozzolino, D. Characterisation and Identification of Individual Intact Goat Muscle Samples (*Capra* sp.) Using a Portable Near-Infrared Spectrometer and Chemometrics. *Foods* **2022**, *11*, 2894. [CrossRef] [PubMed]
7. Cozzolino, D.; Bureš, D.; Hoffman, L.C. Evaluating the Use of a Similarity Index (SI) Combined with near Infrared (NIR) Spectroscopy as Method in Meat Species Authenticity. *Foods* **2023**, *12*, 182. [CrossRef] [PubMed]
8. Luo, X.; Wang, X.; Du, M.; Xu, X. Dispersive Liquid-Liquid Microextraction Followed by HS-SPME for the Determination of Flavor Enhancers in Seafood Using GC-MS. *Foods* **2022**, *11*, 1507. [CrossRef] [PubMed]
9. Ni, D.; Smyth, H.E.; Gidley, M.J.; Cozzolino, D. Predicting Satiety from the Analysis of Human Saliva Using Mid-Infrared Spectroscopy Combined with Chemometrics. *Foods* **2022**, *11*, 711. [CrossRef] [PubMed]
10. Ivanova, M.; Hanganu, A.; Dumitriu, R.; Tociu, M.; Ivanov, G.; Stavarache, C.; Popescu, L.; Ghendov-Mosanu, A.; Sturza, R.; Deleanu, C.; et al. Saponification Value of Fats and Oils as Determined from 1H-NMR Data: The Case of Dairy Fats. *Foods* **2022**, *11*, 1466. [CrossRef] [PubMed]

MDPI

Article

Rapid Food Authentication Using a Portable Laser-Induced Breakdown Spectroscopy System

Xi Wu [1], Sungho Shin [1], Carmen Gondhalekar [1,2], Valery Patsekin [1], Euiwon Bae [3], J. Paul Robinson [1,2] and Bartek Rajwa [4,*]

1 Department of Basic Medical Sciences, Purdue University, West Lafayette, IN 47907, USA
2 Weldon School of Biomedical Engineering, Purdue University, West Lafayette, IN 47907, USA
3 School of Mechanical Engineering, Purdue University, West Lafayette, IN 47907, USA
4 Bindley Bioscience Center, Purdue University, West Lafayette, IN 47907, USA
* Correspondence: brajwa@purdue.edu; Tel.: +1-765-496-1153

Abstract: Laser-induced breakdown spectroscopy (LIBS) is an atomic-emission spectroscopy technique that employs a focused laser beam to produce microplasma. Although LIBS was designed for applications in the field of materials science, it has lately been proposed as a method for the compositional analysis of agricultural goods. We deployed commercial handheld LIBS equipment to illustrate the performance of this promising optical technology in the context of food authentication, as the growing incidence of food fraud necessitates the development of novel portable methods for detection. We focused on regional agricultural commodities such as European Alpine-style cheeses, coffee, spices, balsamic vinegar, and vanilla extracts. Liquid examples, including seven balsamic vinegar products and six representatives of vanilla extract, were measured on a nitrocellulose membrane. No sample preparation was required for solid foods, which consisted of seven brands of coffee beans, sixteen varieties of Alpine-style cheeses, and eight different spices. The pre-processed and standardized LIBS spectra were used to train and test the elastic net-regularized multinomial classifier. The performance of the portable and benchtop LIBS systems was compared and described. The results indicate that field-deployable, portable LIBS devices provide a robust, accurate, and simple-to-use platform for agricultural product verification that requires minimal sample preparation, if any.

Keywords: authentication; LIBS; spectroscopy; food fraud

Citation: Wu, X.; Shin, S.; Gondhalekar, C.; Patsekin, V.; Bae, E.; Robinson, J.P.; Rajwa, B. Rapid Food Authentication Using a Portable Laser-Induced Breakdown Spectroscopy System. *Foods* **2023**, *12*, 402. https://doi.org/10.3390/foods12020402

Academic Editors: Moshe Rosenberg and Daniel Cozzolino

Received: 21 October 2022
Revised: 13 December 2022
Accepted: 5 January 2023
Published: 14 January 2023

1. Introduction

Food fraud, including economically motivated adulteration (EMA), is defined by the US Food and Drug Administration (FDA) as an act in which a valuable ingredient or component of a food product is intentionally omitted, removed, or replaced by a substitute. EMA occurs, as well, when a substance is added to food in order to enhance its appearance, taste, or perceived value [1–3]. Food fraud may involve the deliberate and intentional substitution, addition, tampering, or misrepresentation of food, food ingredients, qualities, or food packaging [2,4].

According to the Food Fraud Database (Decernis LLC, Washington, DC, USA), common examples of affected foods include coffee, cheese, olive oil, herbs and spices, seafood, meat, poultry, alcoholic beverages, honey, fruit and vegetable juices, and cereals. As of 2017, the greatest number of food fraud incidents was associated with dairy products [5–7]. The quality of dairy products in general, and cheeses in particular, was the most frequently reported issue in terms of safety (presence of pathogenic microorganisms), fraud incidences (fraudulent documentation), and adulteration (presence of wood pulp) [7–11]. Many highly valued artisanal cheeses are identified by protected designation of origin (PDO), which helps protect small manufacturers (and local economies) by guaranteeing the authenticity of their products and supporting quality maintenance [12]. Hence, in this study, we selected European Alpine-style cheeses, in addition to coffee, powdered spices, vanilla extract, and balsamic vinegar, to demonstrate the efficacy of our approach [13–16]. A rapidly growing number of reports on food fraud further emphasize the importance of the topic [17].

Rapid classification and authentication of food ensure that fraudulent products do not reach the market or are quickly and efficiently withdrawn. Vibrational spectroscopy, fluorescence spectroscopy, hyperspectral imaging, PCR-based approaches, mass spectrometry, and liquid chromatography are the currently used technologies for detecting food adulterants specifically and food fraud in general [18–23]. Regrettably, each of these approaches requires extensive sample preparation, costly laboratory equipment, highly skilled technicians, and, in some instances, multiple chemical reagents. Regardless of which method is used, there is a considerable time factor associated with the analytical steps.

Laser-induced breakdown spectroscopy (LIBS) has previously been explored as an analytical approach for assessing food integrity [22,24–30], and it is considered to be a promising and exciting method by experts [28,31]. It is a technique that directs a high-energy laser pulse to the surface of a material, resulting in the generation of plasma above this surface and the subsequent emission of optical radiation characteristic of the elements, ions, and molecules that originally comprised the sample [28,32,33]. Analyses of the plasma's optical emission can be used to determine the elemental makeup of the source material [34]. The advantages of LIBS include multi-element detection ability, speed of sampling, and compatibility with a variety of samples (solids, liquids, and gases) [22,33]. In addition, LIBS requires minimal sample preparation and can be used in tandem with other analytical techniques, such as mass spectrometry and Raman spectroscopy [35,36]. LIBS has been used to evaluate milk, infant formula, butter, honey, bakery products, coffee, tea, vegetable oils, water, cereals, flour, potatoes, palm dates, and various types of meat [27,34,37–49]. Moncayo et al. [50] employed LIBS for the authentication of red wines and the localization of their geographic origin. Bilge, et al. [45] discriminated between beef, chicken, and pork meats using LIBS. LIBS was used to identify kudzu powder from different habitats [51], establish the geographical origin of rice [24,52,53], and identify olive oil [54–56].

Herein, the purpose of this study was to determine whether LIBS was a viable choice for identifying food products in various forms (liquid, solid, and powder food samples), using classification models to detect food fraud cases (mislabeling). Two LIBS systems were evaluated to establish the analytical capabilities of LIBS: a benchtop laboratory-based system and a portable device. To our knowledge, this is the first study to use portable LIBS systems for classification analysis of these high-value food goods with the goal of ensuring their authenticity. This is critical since the long-term efficacy of LIBS-based food authentication depends on the availability of portable diagnostic equipment capable of preventing food fraud across the commercial distribution chain, especially for highly valued commodities.

2. Materials and Methods

2.1. Types of Food Samples and Sample Preparation

LIBS is often used on solid samples like metal and plastic that can be recycled. However, food samples in general and liquid food samples in particular present some extra challenges. Because of this, we chose several types of food samples, including liquids, solids, and powders, to represent a wide range of product categories (Table 1).

Table 1. Summary of food samples tested in the study.

Food Forms	Liquid		Solid		Powder
Products	Balsamic vinegar	Vanilla extract	Coffee beans	Cheeses	Spices
Varieties or brands	6	6	7	16	8
Testing methods	NC membrane	NC membrane	Surface shots	Surface shots	Surface shots

2.1.1. Liquid Samples

Balsamic Vinegar

Six types of balsamic vinegar were acquired and tested in the study. These examples were chosen to represent the major brands with distinct protected designations of origin, including three different brands of Modena balsamic vinegar from Italy, barrel-aged balsamic vinegar from Napa Valley (Nap, CA, USA), and Gran Deposito Aceto Balsamico di Modena (Italy), as well as a sample of home-produced barrel-aged balsamic vinegar generously provided by Prof. Andrea Cossarizza (the University of Modena and Reggio Emilia, Italy). A list of the brand names of balsamic vinegar used in the study is provided in Table A1 in Appendix A.

For the measurements of liquid samples in the study, a method utilizing nitrocellulose paper was used. Ten microliters of a sample were deposited onto a 6 × 6-mm nitrocellulose square. Four independent samples of each product were analyzed. There was uneven sample distribution exhibited on the nitrocellulose paper from two samples due to their viscosity. One-to-one dilution with deionized water (DI) was used to resolve it. Samples containing only 10 µL of MilliQ on nitrocellulose squares were used as negative controls. Each nitrocellulose square was measured at different locations 25 times to account for variability and augment the representative dataset.

Vanilla Extracts

A total of six vanilla extract samples were acquired for this study from local stores (West Lafayette, IN). Among them were four vanilla extracts from different geographic locations, represented by different brands, and one vanilla syrup; the remaining one was an imitation vanilla extract composed using artificial flavors. Brand names of the six vanilla products measured in the study are listed in Table A2, Appendix A.

A method similar to that used for measuring the balsamic vinegar (nitrocellulose) was employed for the vanilla extract samples. Briefly, 10 µL of each sample was deposited on a 6 × 6-mm nitrocellulose square and dried at room temperature for 30 min. Each brand was represented by four nitrocellulose-based samples. Due to the high viscosity of the vanilla syrup, one-to-two dilutions with DI water were prepared. As before, 10 µL of DI water on nitrocellulose squares served as the negative control. Each nitrocellulose square was shot 25 times at multiple locations.

2.1.2. Solid Samples

Cheeses

Fifteen types of European Alpine-style cheese purchased from iGourmet, a web-based food delivery service, were shipped as refrigerated 5- to 10-oz. blocks (from 141.75 to 283.5 g). Separately, American Gruyère-style cheese was purchased from a local Kroger supermarket. This product is referred to as Wisconsin Gruyère cheese in the study. A total of 16 types of cheeses are listed in Table A3, Appendix A.

Cheeses were stored at 4 ± 1 °C until analysis. Approximately 1 cm of the outside of the cheese block was cut and discarded to prevent the use of dried material. For LIBS measurement, cheese samples were cut into rectangular slices of uniform thickness (approximately 10 mm wide, 10 mm long, and 2 mm thick) using a stainless-steel blade. For each time point, four replicate specimens were cut from each type of cheese block. The blade was rinsed and cleaned with ethanol and dried between each cut of the same cheese and between each cut of different cheeses.

Water activity (a_w) was determined for the sixteen Alpine-style cheeses every two weeks for 42 days of storage in a refrigerator. The purpose was to establish data regarding the impact of storage on the LIBS-based product classification. In short, grated cheese samples (0.5 g) were placed in plastic dishes, covered, stored at 4 °C, and assayed in duplicate at 25 °C on an AquaLab 4TE Dew Point Water Activity Meter (AquaLab, Pullman, WA, USA). The precise dewpoint temperature of the sample was established by an infrared beam focused on a small mirror. The temperature at the dewpoint was then converted into water activity. Prior to analysis, the machine was calibrated using a certified AQUA LAB standard (Lot no. 20805392, 0.920 a_w NaCl, 2.33 mol/Kg in H_2O). The a_w of the cheese was measured at 0 (T1), 14 (T2), 28 (T3), and 42 (T4) days, along with the LIBS measurement. The a_w data were expressed as the mean of three repetitions in three independent measurements. Utilizing commercially accessible software, data were analyzed using 2-way ANOVA and Tukey's multiple comparisons test (OriginPro, OriginLab Corporation, Northampton, MA, USA). Comparisons were considered significantly different at a p-value < 0.05.

Coffee Beans

In this study, seven varieties of coffee were tested directly without the need for grinding or milling. Whole coffee beans were stored in the original sealed package until the test and resealed after use. The names of the coffee varieties tested in the study are listed in Table A4, Appendix A.

Four randomly selected coffee beans of each type were measured from both the front and back sides. To avoid additional variability caused by the movement of the beans when hit by the laser, the coffee beans were fastened with tape to a sample holder. The location of the beans was adjusted for multiple LIBS interrogations to cover as much area on the bean surface as possible.

2.1.3. Powdered Food Samples

Spices

Six different types of spices were chosen and purchased from the retail outlets. Table A5 in Appendix A provides the brand names of the spices evaluated in the study.

Most of the ground spices used in this study are fine powders, although the classic nutmeg is roughly milled powder. The red pepper comes as flakes, which splash easily when hit by laser shots. Therefore, we employed a sample holder when performing the measurements.

2.2. Benchtop and Handheld LIBS Systems Setup

The custom-built benchtop LIBS system is shown in Figure 1a and consists of a Nano SG 150-10 pulsed Nd:YAG laser (Litron Lasers, Bozeman, MT, USA). The laser had a pulse width of 4 ns; a pulse energy of 62 mJ was used in this study. The ablation laser's spot size was approximately 700 µm. Details on the optics used to direct the alignment and the ablation laser beams were described previously [57,58]. Emissions were detected by an AvaSpec-Mini-VIS-OEM spectrometer (Avantes, Apeldoorn, the Netherlands), which has a 350–600-nm spectral range with 0.33-nm resolution. Target samples were placed on a motorized XYZ stage. The stage height was adjusted so that the crosshairs of the two lasers assisting in sample positioning were visible at the surface of the samples. A digital delay pulse generator controlled the triggering of the ablation laser, motorized stage, and spectrometer. The delay between the ablation pulse and spectrometer data acquisition was 1.17 µs.

The Z-300 LIBS Analyzer (SciAps, Inc., Boston, MA, USA) is a commercially available handheld LIBS system. The laser, spectrometer, optics, argon gas cartridge, electronics, and control module were housed in a gun-shaped enclosure, as illustrated in Figure 1b. Measurements were performed when the sample window (3 cm by 3 cm) was covered with samples, followed by laser activation. The LIBS analyzer uses a pulsed laser, 5–6 mJ/pulse, and 1- to 2-ns pulse width. The laser spot size was 100 µm. The spectral range was approximately 190–950 nm. The settings for rastering location and repetition rate were controlled in the Profile Builder software (SciAps, Inc.) as needed.

All measurements were taken at 25 different locations across a 5×5 rastering array of four different specimens representing each individual food product. The measurements of cheeses were repeated at multiple time points (Figure 2). Each spot was ablated with a single laser shot. Accordingly, 100 spectra per food type per time point were analyzed for classification. LIBS measurements were performed using both benchtop and handheld systems for each type of food sample involved in the study.

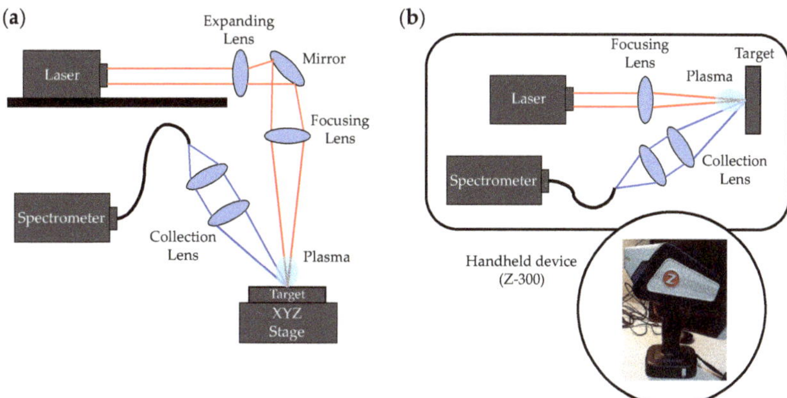

Figure 1. Schematic diagram of LIBS system setup; (**a**) benchtop system and (**b**) handheld system.

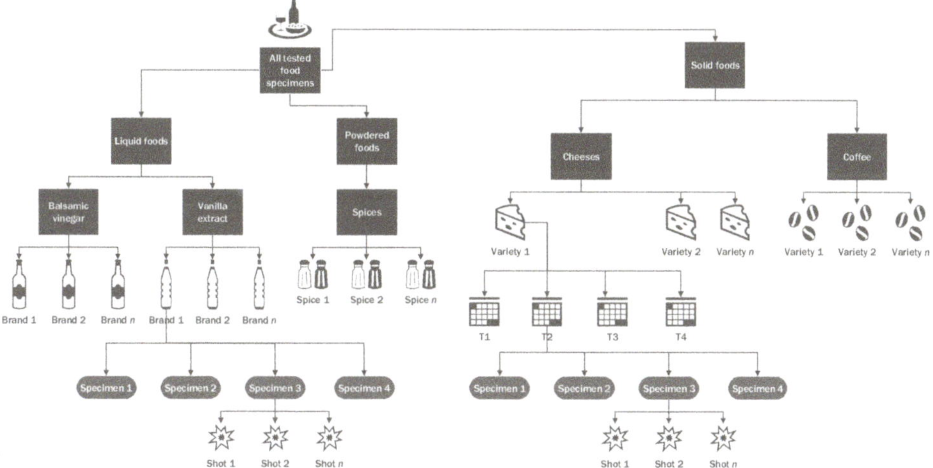

Figure 2. Diagram illustrating the variety of food examples and the testing procedures employed in the presented experiments. Each food product was represented by multiple specimens, each of which was interrogated repeatedly by LIBS. Please note that only cheeses were sampled at multiple time intervals.

2.3. Classification Procedures

Raw spectra were filtered to eliminate low signal-to-noise instances due to faulty sample positioning or similar technical problems. Spectral normalization and a median filter were applied to reduce the effects of variations in emission intensity coming from plasma fluctuations. Subsequently, every spectral feature was used in multiple ANOVA models as a dependent variable in order to select the features associated with large effect sizes (η^2) [59]. This was followed by the training of a regularized multinomial logistic regression elastic net model (ENET), which provides multivariate feature selection as well as classification (prediction) [60,61]. ENET combines LASSO and ridge regression techniques. Although the use of the ENET approach in LIBS data analysis has been reported before [62], despite its advantages, it is still a very uncommon method for this field, which traditionally relies on well-established chemometric techniques such as PLS-DA [63–66]. Importantly, in the n≪p setting, it retains the sparse features of LASSO regression and the stability of ridge regression. Note that the number of selected features per food type could differ for each ENET model. The ENET prediction quality was evaluated using 10-fold cross-validation.

3. Results

3.1. LIBS Measurements

Table 1 summarizes all the food products measured in the study. We tested three different forms of high-value regional foods (liquid, solid/semi-solid, and powder) by both benchtop and handheld LIBS, including 16 hard cheeses, seven coffee varieties, six vanilla/vanillin extracts, and six different powdered spices. Additionally, we monitored changes in the water activity of the cheese samples at four sampling time points. It is known that water-activity measurement is an important method for predicting the shelf life of food products. By measuring and controlling the water activity of foodstuffs, it is possible to monitor and maintain the physical stability of foods and optimize their physical properties. Therefore, the water activity of cheeses is an indicator informing us about the shelf-life status of the product. Figure 3 illustrates the evolution of water activity in the test cheeses during a period of refrigerated storage.

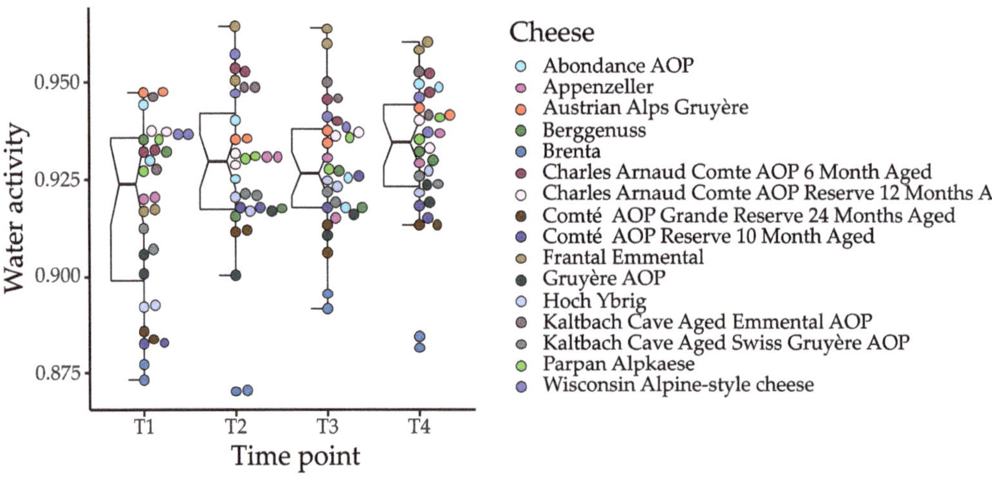

Figure 3. Changes in water activity in 16 types of tested cheeses over six weeks of refrigerated storage measured at four time-points.

All the food samples were measured by the benchtop LIBS system covering a spectral window from 200 to 600 nm. The corresponding data obtained from the handheld LIBS device covered a spectral range of 190 to 950 nm. The typical LIBS spectra of (a) coffee bean, (b) vanilla extract, (c) balsamic vinegar, and (d) spice samples, measured using benchtop and handheld LIBS systems, are shown in Figures 4 and 5, respectively. The spectra of each food category represent an average of all the measurements. For example, Figure 4b is an averaged spectrum of 600 (six vanilla extracts × 100 spectra/vanilla extract) measurements. The data in Figure 5 are spectral results obtained after automatic data processing executed in the handheld device, whereas Figure 4 represents the raw data from the benchtop system. The main emission lines from the essential elements for food analysis, selected as inputs of ENET, have been labeled in Figures 4a and 5a. The detected elemental emission bands are identified with the aid of the spectroscopic data included in the NIST Atomic Spectra Database [67]. CN band, Ca ionic, Ca atomic, C_2 band, P ionic, and Na atomic peaks, which are dominantly detected in biomaterials, can be seen in Figure 4.

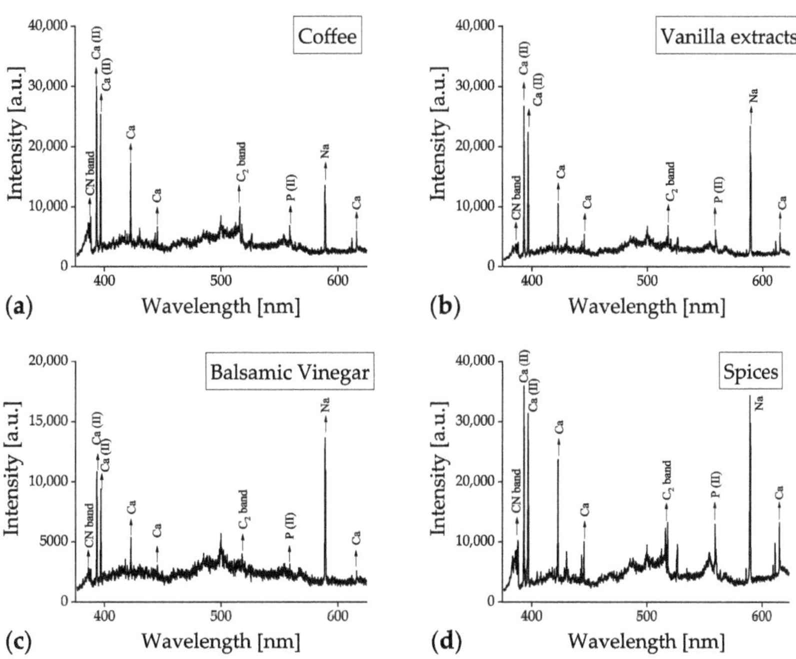

Figure 4. Averaged raw LIBS spectra of (**a**) coffee, (**b**) vanilla extract, (**c**) balsamic vinegar, and (**d**) spice samples collected using the benchtop LIBS system.

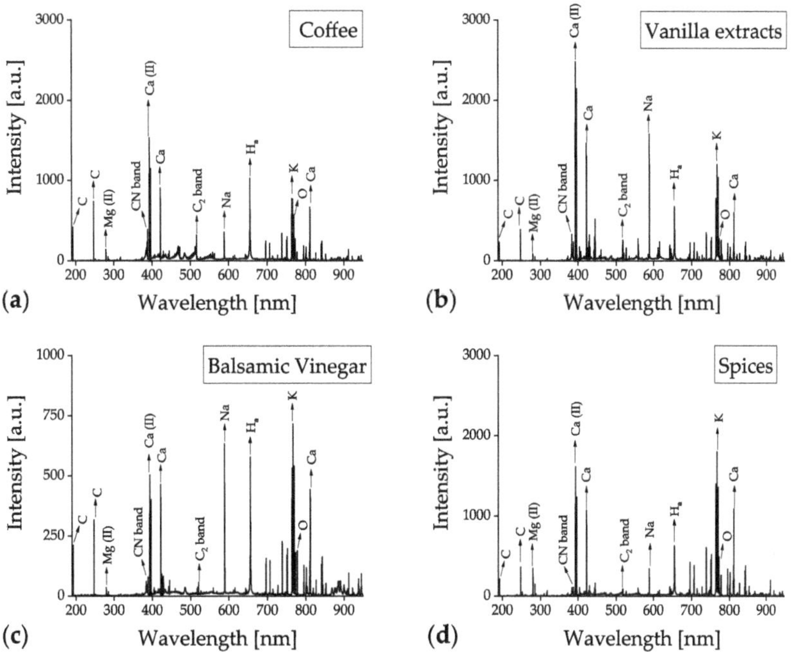

Figure 5. Averaged raw LIBS spectra of (**a**) coffee, (**b**) vanilla extract, (**c**) balsamic vinegar, and (**d**) spice samples collected using the handheld LIBS system.

Although there was a minor difference in peak values depending on the food products, the same emission peaks were found in all the tested food samples. Similarly, there were only minor differences in the handheld LIBS results, as shown in Figure 5. However, additional peaks, such as C, Mg, H, K, and O peaks, were detected owing to the broader spectral range (190–950 nm) of the handheld device. This broader spectral range contributed to improving the classification accuracy of the coffee bean, vanilla extract, and balsamic vinegar samples.

Figures 6 and 7 show the averaged LIBS spectra of the cheese samples, measured using the benchtop and handheld LIBS systems at four different time points. Note that each spectrum is an average of 1600 (16 cheese types × 100 spectra/cheese type) measurements under the same conditions. The measurements were conducted every 14 days. The cheese specimens were instantly stored in a vacuum pack and refrigerator after each measurement. Emissions of the identical elemental components in various LIBS spectral fingerprints of the cheese samples led to a significant degree of resemblance. Some minor differences in peak intensities appeared at different time points. As an example of changes over time, Table 2 compares the integrated peak intensity of Na I 589.0 nm in Frantal Emmental Cheese (C10) at each sampling time point. Integrated peak intensity was calculated by integrating the peak area study after sum-to-one normalization. It was shown that the averaged integrated intensities of the Na I emission peak were similar at four different sampling time points, implying relatively uniform product preservation within time periods.

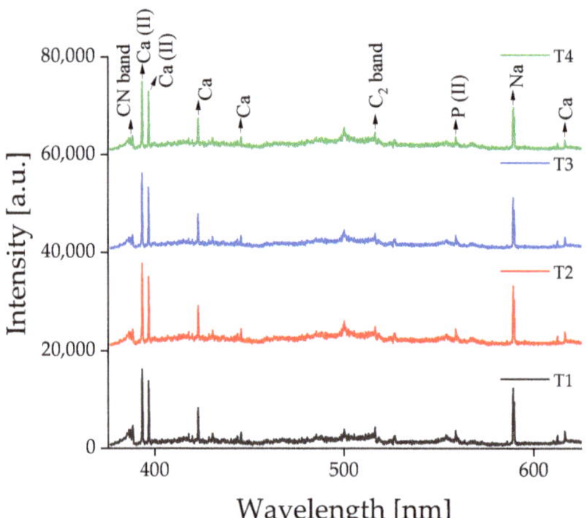

Figure 6. Averaged raw LIBS spectra of cheese samples measured on four different dates using the benchtop LIBS system. Note that every measurement was conducted every two weeks.

Table 2. The averaged integrated intensity of emission peak Na I 589.0 nm at four different sampling time points in Frantal Emmental Cheese (C10). The values in parentheses represent the relative standard deviation (RSD).

Time Point	Benchtop LIBS	Handheld LIBS
T1	0.0060 (17.4%)	0.0071 (11.3%)
T2	0.0062 (10.6%)	0.0063 (15.3%)
T3	0.0056 (10.0%)	0.0067 (16.3%)
T4	0.0054 (18.6%)	0.0074 (18.3%)

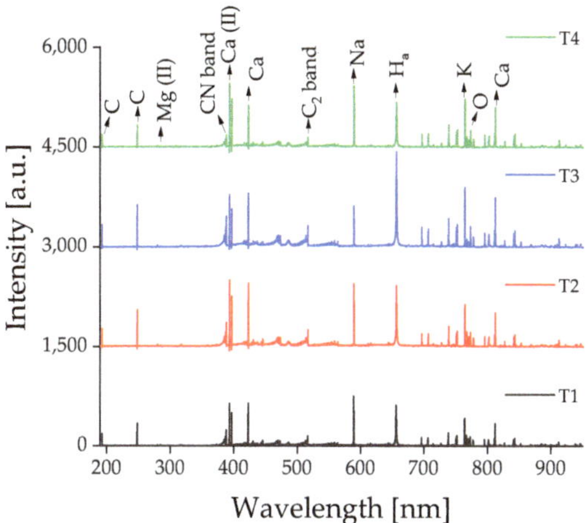

Figure 7. Averaged raw LIBS spectra of cheese samples measured on four different dates using the handheld LIBS system. Note that every measurement was conducted every two weeks.

3.2. Classification Using the Elastic Net Approach

Table 3 reports the ENET classification accuracy of five different food products measured using the benchtop LIBS system and the handheld LIBS system. The training (and accuracy evaluation) was performed via 10-fold cross-validation. As can be seen in the tables, cheese samples were sampled and measured by two LIBS systems at four time points. Thus, separate classifiers were developed and applied to the dataset containing measurements from each of the four time points. As mentioned before, prior to the algorithmic training, univariate feature selection via ANOVA was applied to the data to minimize the subsequent training time. The accuracy of the model was found to be excellent, reaching 94.5 ± 1.51% for vanilla extract and 99.30 ± 0.70% for spices in the benchtop system, and 92.70 ± 2.30% for coffee beans, 98.30 ± 0.69% for vanilla extract, and 90.80 ± 1.88% for balsamic vinegar in the handheld system.

Table 3. ENET classification accuracy of five different food products measured by the benchtop and handheld LIBS systems at four different time points.

Food Products	Classifier Accuracy	
	Benchtop LIBS	**Handheld LIBS**
16 cheeses		
T1	85.80 ± 1.57%	81.20 ± 1.51%
T2	82.20 ± 1.53%	83.00 ± 1.34%
T3	87.60 ± 1.99%	84.70 ± 1.79%
T4	84.10 ± 1.93%	84.20 ± 1.71%
6 coffee varieties	85.00 ± 1.94%	92.70 ± 2.30%
6 vanilla extracts	94.50 ± 1.51%	98.30 ± 0.69%
6 balsamic vinegars	88.20 ± 2.10%	90.80 ± 1.88%
8 powdered spices	99.30 ± 0.70%	84.50 ± 1.94%

The classification of coffees and balsamic vinegar showed slightly lower accuracy in the benchtop system compared to the handheld system. This suggests that the broad spectral range of the handheld system may be the most dominant factor in the classification of coffee beans and balsamic vinegar using LIBS. However, the classification accuracy of spices in powder form was lower using the handheld system, pointing to the spectral resolution as the decisive factor. Additional studies are required to evaluate these types of samples further, particularly with respect to the preparation methods for powders. The test results for vanilla extracts show comparable classification accuracy in both LIBS systems.

The classification performance for cheese samples measured at different storage time points was also assessed. There were no observable differences in the measurements obtained during different periods. The classification accuracy of those measurements did not present significant differences either. Note that every three sample replicates were averaged and analyzed to establish the classification performance results. Slightly higher classification accuracy of cheese samples was shown in the benchtop system than in the handheld device. It is likely that more sensitive detection in the visible and near-visible range (350–650-nm wavelength) could be the critical factor for classifying cheeses using LIBS.

3.3. Food Fraud Detection

In the final step of our study, we simulated two realistic food-fraud detection scenarios in which a specific sample with a different origin and/or composition than the rest of the set was to be identified and detected. In the first simulation, we aimed to identify Wisconsin Gruyère-style cheese manufactured in the USA from pasteurized milk. In the second scenario, we attempted to identify imitation vanilla taste (vanillin) among natural vanilla extracts. In the first scenario, we envisioned three classes (unpasteurized European cheeses branded as "Gruyère" vs. other unpasteurized European Alpine-style cheeses vs. Wisconsin Gruyère-type cheese produced from pasteurized milk), whereas, in the second scenario, there are only two classes (real vanilla extract vs. imitation vanilla flavor). We used multiple repeated independent instances of 5×2 cross-validation runs to evaluate the system. For the cheese detection scenario, the accuracies of the benchtop ($90.17 \pm 1.04\%$) and the portable platforms ($90.95 \pm 1.05\%$) were virtually identical (see Table A6). Similarly, the benchtop and the portable systems operated equally well in detecting the imitation vanilla (99.66 ± 0.47 and $99.38 \pm 0.58\%$, respectively). See Table A7 in Appendix A for the result of the individual classification runs.

4. Discussion

4.1. Sample Preparation

Solid specimens were successfully analyzed without any processing. Grinding samples into powder and pressing them into a pellet is a popular preparation method for solid foods [68,69]. For instance, Iqbal et al. [70] reported that samples were finely powdered and vacuum-dried at 370 K for 10 h. The sample was then compressed for 20 min at 30 T hydraulic pressure into pellets that were 3 mm thick and 1.3 cm in diameter. However, the preparation of pellets or tablets is an important limiting factor and cannot be easily used for in-situ analysis. In contrast, in our experiments, solid food samples like Alpine cheeses and coffee were tested without any preparation. The samples were immobilized for an easy location adjustment to ensure coverage of the whole sample surface by laser shots during the collection of complete elemental profiles.

Regarding measurement preparation for powders (spices), we utilized a custom sample holder to confine the samples. To overcome blowing off and scattering during laser-matter interaction, a layer of powdered material was applied to a double-sided piece of tape that covered and adhered to the bottom of the sample holder.

To prevent splashing and the formation of surface ripples caused by the shock wave of LIBS, as well as to achieve a lower limit of detection, better repeatability, and greater sensitivity when working with liquid food samples, the formation of a gel using commercial collagen is commonly performed, followed by drying in an air-assisted oven [50,71,72]. However, the dry gel emission signals cannot be simply subtracted, and additional chemometric spectral treatments are necessary. In our study, we employed a nitrocellulose paper-based sample-preparation approach that is highly compatible with liquid food samples owing to its porous structure, hydrophilic property, and minimal effect on the sample spectra. This approach has been successfully used by other researchers when utilizing

LIBS to measure the presence of metals in water or oil [73–76]. Moreover, this method is simpler and more efficient than the commonly used gel-formation technique [77]. The characteristic peaks of the nitrocellulose membrane do not interfere with the elemental profiles of foods and can be easily distinguished from the LIBS spectral matrix. This is the first report on the use of nitrocellulose membranes with LIBS for the classification of liquid food samples. Compared to the commonly used methods, our approach requires little or no sample preparation. It is simple, rapid, and cost-effective. Consequently, it is more practical and compatible with envisioned usage scenarios involving wholesalers, food inspectors, and customs officers that examine traded agricultural products. However, we must stress that the viability of using nitrocellulose paper may depend on the viscosity of the sample. We have not tested a sufficient range of liquid products to endorse this method unreservedly.

4.2. Water Activity

Most of the 16 types of cheese showed a small but statistically significant difference in water-activity values between the beginning of storage and 42 days later. However, despite these small changes in water activity, the classification of cheeses with LIBS systems remained stable and robust. Interestingly, one recent LIBS application was to measure the moisture content in cheese, using oxygen emission normalized by CN emission as the indicator [78]. Another study performed by Ayvaz et al. [79] investigated the potential of using LIBS with partial least squares regression to determine the chemical quality-control parameters for cheese samples, such as moisture, dry matter, salt, total ash, total protein, and pH. In general, our results indicate that small variations in a_w are unlikely to be limiting factors for the use of LIBS in authentication, provided that the classification system is paired with an appropriate feature-selection strategy.

4.3. Spectral Classification

As anticipated, the LIBS spectra of all the analyzed food items exhibit remarkably similar spectral characteristics due to their comparable elemental composition. Clearly, the significant resemblance between these spectra makes their classification challenging, at least visually. For the differentiation and classification of food samples based on their LIBS spectra, it is therefore required to employ statistical machine-learning approaches.

We chose ENET as the primary tool for analyzing LIBS spectra due to its embedded feature selection capability, which is crucial given the usage of high-resolution spectra and a restricted number of food samples [60,61]. The ENET method classifies products using LIBS while identifying the most relevant chemical constituents that support the classification results. It is important to note, however, that features identified by ML algorithms may not always represent identifiable elemental peaks and may also come from "background". Matrix effects play a big role in how complex samples (like food) are measured by spectroscopy, and multivariate approaches may exploit the matrix effects when fingerprinting is performed [80].

To the best of our knowledge, relatively few published studies apply LIBS supported by machine-learning algorithms to discriminate/classify food samples based on their geographical origins or detection of adulteration. As for liquid food samples, three research reports have indicated that LIBS techniques paired with machine-learning approaches were employed with success for the discrimination/classification of several olive oils according to their acidity and geographical origin [54–56]. The olive oils tested in these studies are distinct in geographical origin and oil quality, i.e., extra virgin olive oil quality or typical commercial edible oils. Oil samples were placed in shallow, uncovered glass Petri plates such that a focused laser beam could reach their free surface to generate plasma. In these studies, classification accuracy rates of more than 90% were achieved, indicating the promise of this method. Considering the limitations and difficulties of working with aqueous samples, researchers developed a liquid-to-solid transformation of red wine using a dry collagen gel to increase the analytical performance. The LIBS technique combined with neural networks provided a classification procedure for the quality control of red wines with PDO [50]. Furthermore, the identification of milk fraud, as well as the adulteration ratios, were reported using LIBS coupled with visual clustering following principal component analysis (PCA) [29].

Previous studies reported using the combination of LIBS and chemometric and/or machine-learning methods to identify coffee varieties [16] and detect adulteration of wheat, corn, and chickpeas in Arabica coffee [68]. The samples were ground and pressed into pellets for LIBS measurements. Zhang et al. tested multiple classifiers (including support vector machines, neural networks, and partial least squares (PLS) regression), some of which provided an accuracy of around 80% [16]. In our study, we achieved a higher classification accuracy by employing the elastic net approach. In the other study, all major and minor elemental composition differences present in the LIBS spectra of coffee were identified using traditional chemometric techniques such as PCA and PLS [81]. In contrast, in our study, the most critical spectral features associated with elemental differences were identified using the embedded feature selection ability of the ENET model. These findings confirmed that the combination of LIBS and the ENET classifier has the potential to be used as a routine technique for determining coffee bean authenticity and detecting adulteration. It is becoming increasingly important to employ chemometrics and machine-learning methods in food authentication systems [82–84]. The fact that ENET allows for simultaneous feature selection (providing insights into the elemental composition), as well as classification, demonstrated that it is a method exceptionally well-suited for this food analysis task.

As far as we know, this study is the first time that LIBS and chemometric methods were used together to classify 16 types of cheese. The results showed that this combination could be a useful and practical way to find food fraud in cheese products without a lot of sample preparation. Also, this is the first study to utilize LIBS assisted by machine-learning methods to efficiently classify powdered spices using direct analysis, i.e., without making pellets. Thus, our results demonstrated that LIBS, aided by suitable statistical methods, can be an effective technique for verifying the quality and safety of spices and similar powdered products.

It is astonishing that there are discernible spectral differences between closely related cheeses. One probable explanation is that artisanal Alpine-style cheeses are produced seasonally in particular regions, and the bacteria responsible for cheese ripening and maturation are distinctively associated with geographical location and changing seasons [85–89].

Even though our classification experiments show a remarkably high degree of accuracy, it is important to note a critical limitation. For each example presented, the tests assume a supervised learning environment with an exhaustively defined training set. In other words, we assume that all classes are known beforehand (including the classes describing possibly fraudulent or inferior products). This cannot be guaranteed in many instances, resulting in the so-called non-exhaustive learning problem, which necessitates simultaneous class discovery and classification [90]. We plan to address this issue in future research using our prior experience with non-exhaustive training sets, such as those emerging in food safety applications [91].

5. Conclusions

The LIBS technique, paired with supervised statistical learning methods, has been evaluated in real-world applications as a rapid and robust classifier of high-value food items based on their distinctive spectral fingerprints. This study aimed to demonstrate that an existing field-deployable LIBS device originally built for material science applications may provide a rapid, easy, and inexpensive authentication platform for agricultural products where minimal or no sample preparation is required. To achieve this purpose, our study utilized new, easy, and cost-effective sample preparation techniques for liquid and powdered food samples. Utilizing nitrocellulose paper for liquid food samples improved the quality of the spectra and allowed us to avoid the typical sample splashing caused by LIBS-generated shock waves. The LIBS signal of nitrocellulose paper is readily distinguished from the spectra of tested food samples. It has also been demonstrated that accurate analysis of solid foods such as cheeses and entire coffee beans may be performed using LIBS without any sample preparation.

Overall, the results point to the feasibility of rapid identification of various high-value foods by LIBS accompanied by supervised classification methods, using not only lab-based benchtop instruments but also portable, field-deployable units.

Author Contributions: Conceptualization, B.R.; methodology, B.R.; software, B.R. and V.P.; instrumentation, E.B. and J.P.R.; data acquisition, C.G., X.W. and S.S.; statistical analysis, S.S. and B.R.; resources, J.P.R.; data curation, S.S.; writing—original draft preparation, X.W.; writing—review and editing, B.R., J.P.R., E.B. and S.S.; funding acquisition, B.R., E.B. and J.P.R. All authors have read and agreed to the published version of the manuscript.

Funding: This research represents the subproject "Development of machine learning tools for LIBS food fingerprinting and classification" and is a part of the multi-investigator "Development of Innovative Technologies and Strategies to Mitigate Biological, Chemical, Physical, and Environmental Threats Food Safety" project supported by the Center for Food Safety Engineering at Purdue University, funded by the U.S. Department of Agriculture, Agricultural Research Service, under Agreement No. 59-8072-1-002. Opinions, findings, conclusions, or recommendations expressed in this publication are those of the authors and do not reflect the view of the U.S. Department of Agriculture.

Data Availability Statement: Not applicable.

Acknowledgments: We thank Lisa J. Mauer for her help with water-activity measurements.

Conflicts of Interest: The authors declare no conflict of interest.

Appendix A

Additional tables containing information about all the tested food samples and the detailed results of the food fraud simulation study are described in Section 3.3.

Table A1. Tested balsamic vinegar samples.

Code	Balsamic Vinegar Samples
B1	Balsamic Vinegar of Modena
B2	Balsamic Vinegar of Modena (Colavita)
B3	Barrel-aged Balsamic Vinegar (Napa Valley Harvest)
B4	Gran Deposito Aceto Balsamico di Modena (Giuseppe Giusti)
B5	Gold Quality Balsamic Vinegar of Modena (Trader Joe's)
B6	Prof. Andrea Cossarizza's private collection balsamic vinegar

Table A2. Tested vanilla samples.

Code	Vanilla Samples
V1	Pure vanilla extract (Kroger, Cincinnati, OH)
V2	Imitation vanilla flavor (Kroger, Cincinnati, OH)
V3	Pure vanilla extract (McCormick & Company, Baltimore, MD)
V4	San Luis Rey pure vanilla (La Vencedora e Hijos SA de CV, San Luis Potosi, Mexico
V5	Vanilla syrup (1883 Maison Routin, La Motte Servolex, France)
V6	Simple Truth Madagascar vanilla extract (Kroger, Cincinnati, OH)

Table A3. Tested Alpine-style cheese samples.

Code	Alpine-Style Cheese Samples
C1	Abondance AOP
C2	Appenzeller
C3	Austrian Alps Gruyère
C4	Berggenuss
C5	Brenta
C6	Charles Arnaud Comté AOP 6 Month Aged
C7	Charles Arnaud Comté AOP Reserve 12 Months Aged
C8	Comté AOP Grande Reserve 24 Months Aged
C9	Comté AOP Reserve 10 Month Aged
C10	Frantal Emmental
C11	Gruyère AOP
C12	Hoch Ybrig
C13	Kaltbach Cave Aged Emmental AOP
C14	Kaltbach Cave Aged Swiss Gruyère AOP
C15	Parpan Alpkaese
C16	Wisconsin Gruyère Alpine-Style Cheese

Table A4. Tested coffee samples.

Code	Coffee Samples
F1	Italian Dark Roast (OLDE Brooklyn Coffee, Brooklyn, NY)
F2	Guatemalan Antigua Blend (Copper Moon Coffee, Lafayette, IN)
F3	Lavazza Super Crema (Luigi Lavazza SpA, New York, NY)
F4	Despierta tus Sentidos (Nespresso USA Inc., Long Island City, NY)
F5	Café Cubano Roast (Mayorga Organics, Rockville, MD)
F6	Artisan Blend (Koffee Kult, Hollywood, FL)
F7	Shot Tower Espresso (Verena Street Coffee Co., Dubuque, IA)

Table A5. Tested spices samples.

Code	Spices Samples
S1	East Indian ground nutmeg (McCormick & Company, Baltimore, MD)
S2	Classic ground nutmeg (McCormick & Company, Baltimore, MD
S3	Ground mustard (Kroger, Cincinnati, OH)
S4	Smidge & Spoon crushed red pepper (Kroger, Cincinnati, OH)
S5	Cayenne pepper (Spice Islands, Ankeny, IA)
S6	Ground cumin (McCormick & Company, Baltimore, MD)
S7	Private Selection ground cumin (Kroger, Cincinnati, OH)
S8	Simple Truth organic ground turmeric (Kroger, Cincinnati, OH)

Table A6. Result of testing three cheese categories (Alpine-style cheeses identified as "Gruyère" manufactured from unpasteurized milk, other Alpine-style cheese produced from unpasteurized milk, Wisconsin Alpine-style cheese produced from pasteurized milk). The table reports 10 independent 5×2 cross-validation runs.

Experiment Run	Accuracy [%]	
	Benchtop	Handheld
1	89.1	91.3
2	91.5	92.3
3	89.4	89.4
4	90.5	91.8
5	90.8	89.9
6	89.8	90.8

Table A6. *Cont.*

Experiment Run	Accuracy [%]	
	Benchtop	Handheld
7	91.6	91.3
8	90.8	89.6
9	88.4	92.3
10	89.8	90.8
	90.17 (1.04)	90.95 (1.05)

Table A7. Result of detecting imitation vanilla (vanillin) among real vanilla extracts. The table reports 10 independent 5 × 2 cross-validation runs.

Experiment Run	Accuracy [%]	
	Benchtop	Handheld
1	99.4	98.8
2	100	98.8
3	100	100
4	98.9	100
5	100	99.0
6	99.4	98.6
7	100	99.2
8	100	100
9	98.9	99.4
10	100	100
	99.66 (0.47)	99.38 (0.58)

References

1. Moyer, D.C.; DeVries, J.W.; Spink, J. The economics of a food fraud incident—Case studies and examples including melamine in wheat gluten. *Food Control* **2017**, *71*, 358–364. [CrossRef]
2. Spink, J.; Moyer, D.C. Defining the public health threat of food fraud. *J. Food Sci.* **2011**, *76*, R157–R163. [CrossRef] [PubMed]
3. Spink, J.; Bedard, B.; Keogh, J.; Moyer, D.C.; Scimeca, J.; Vasan, A. International survey of food fraud and related terminology: Preliminary results and discussion. *J. Food Sci.* **2019**, *84*, 2705–2718. [CrossRef]
4. Manning, L.; Soon, J.M. Food safety, food fraud, and food defense: A fast evolving literature. *J. Food Sci.* **2016**, *81*, R823–R834. [CrossRef]
5. Decernis. Food Fraud Database. Available online: https://decernis.com/products/food-fraud-database/ (accessed on 12 December 2022).
6. Manning, L.; Soon, J.M. Food fraud vulnerability assessment: Reliable data sources and effective assessment approaches. *Trends Food Sci. Technol.* **2019**, *91*, 159–168. [CrossRef]
7. Montgomery, H.; Haughey, S.A.; Elliott, C.T. Recent food safety and fraud issues within the dairy supply chain (2015–2019). *Glob. Food Secur.* **2020**, *26*, 100447. [CrossRef]
8. Popping, B.; De Dominicis, E.; Dante, M.; Nocetti, M. Identification of the geographic origin of Parmigiano Reggiano (P.D.O.) cheeses deploying non-targeted mass spectrometry and chemometrics. *Foods* **2017**, *6*, 13. [CrossRef] [PubMed]
9. Gimonkar, S.; Van Fleet, E.; Boys, K.A. Chapter 13—Dairy product fraud. In *Food Fraud*; Hellberg, R.S., Everstine, K., Sklare, S.A., Eds.; Academic Press: Cambridge, MA, USA, 2021; pp. 249–279.
10. Cardin, M.; Cardazzo, B.; Mounier, J.; Novelli, E.; Coton, M.; Coton, E. Authenticity and Typicity of Traditional Cheeses: A Review on Geographical Origin Authentication Methods. *Foods* **2022**, *11*, 3379. [CrossRef]
11. Amilien, V.; Moity-Maïzi, P. Controversy and sustainability for geographical indications and localized agro-food systems: Thinking about a dynamic link. *Br. Food J.* **2019**, *121*, 2981–2994. [CrossRef]
12. European Commission. European Commission Food Quality Schemes Explained. Available online: https://ec.europa.eu/info/food-farming-fisheries/food-safety-and-quality/certification/quality-labels/quality-schemes-explained_en (accessed on 12 December 2022).
13. Ferreira, T.; Galluzzi, L.; de Paulis, T.; Farah, A. Three centuries on the science of coffee authenticity control. *Food Res. Int.* **2021**, *149*, 110690. [CrossRef]
14. Galvin-King, P.; Haughey, S.A.; Elliott, C.T. Herb and spice fraud; the drivers, challenges and detection. *Food Control* **2018**, *88*, 85–97. [CrossRef]
15. Silvis, I.C.J.; van Ruth, S.M.; van der Fels-Klerx, H.J.; Luning, P.A. Assessment of food fraud vulnerability in the spices chain: An explorative study. *Food Control* **2017**, *81*, 80–87. [CrossRef]

16. Zhang, C.; Shen, T.; Liu, F.; He, Y. Identification of coffee varieties using laser-induced breakdown spectroscopy and chemometrics. *Sensors* **2018**, *18*, 95. [CrossRef] [PubMed]
17. Gussow, K.E.; Mariët, A. The scope of food fraud revisited. *Crime Law Soc. Chang.* **2022**, *78*, 621–642. [CrossRef]
18. Ellis, D.I.; Muhamadali, H.; Haughey, S.A.; Elliott, C.T.; Goodacre, R. Point-and-shoot: Rapid quantitative detection methods for on-site food fraud analysis—Moving out of the laboratory and into the food supply chain. *Anal. Methods* **2015**, *7*, 9401–9414. [CrossRef]
19. Perestam, A.T.; Fujisaki, K.K.; Nava, O.; Hellberg, R.S. Comparison of real-time PCR and ELISA-based methods for the detection of beef and pork in processed meat products. *Food Control* **2017**, *71*, 346–352. [CrossRef]
20. López-Maestresalas, A.; Insausti, K.; Jarén, C.; Pérez-Roncal, C.; Urrutia, O.; Beriain, M.J.; Arazuri, S. Detection of minced lamb and beef fraud using NIR spectroscopy. *Food Control* **2019**, *98*, 465–473. [CrossRef]
21. Nunes, K.M.; Andrade, M.V.O.; Almeida, M.R.; Fantini, C.; Sena, M.M. Raman spectroscopy and discriminant analysis applied to the detection of frauds in bovine meat by the addition of salts and carrageenan. *Microchem. J.* **2019**, *147*, 582–589. [CrossRef]
22. Hassoun, A.; Måge, I.; Schmidt, W.F.; Temiz, H.T.; Li, L.; Kim, H.-Y.; Nilsen, H.; Biancolillo, A.; Aït-Kaddour, A.; Sikorski, M.; et al. Fraud in animal origin food products: Advances in emerging spectroscopic detection methods over the past five years. *Foods* **2020**, *9*, 1069. [CrossRef]
23. Hebling e Tavares, J.P.; da Silva Medeiros, M.L.; Barbin, D.F. Near-infrared techniques for fraud detection in dairy products: A review. *J. Food Sci.* **2022**, *87*, 1943–1960. [CrossRef]
24. Kim, G.; Kwak, J.; Choi, J.; Park, K. Detection of nutrient elements and contamination by pesticides in spinach and rice samples using laser-induced breakdown spectroscopy (LIBS). *J. Agric. Food Chem.* **2012**, *60*, 718–724. [CrossRef]
25. Multari, R.A.; Cremers, D.A.; Dupre, J.A.M.; Gustafson, J.E. Detection of biological contaminants on foods and food surfaces using laser-induced breakdown spectroscopy (LIBS). *J. Agric. Food Chem.* **2013**, *61*, 8687–8694. [CrossRef] [PubMed]
26. Multari, R.A.; Cremers, D.A.; Scott, T.; Kendrick, P. Detection of pesticides and dioxins in tissue fats and rendering oils using laser-induced breakdown spectroscopy (LIBS). *J. Agric. Food Chem.* **2013**, *61*, 2348–2357. [CrossRef]
27. Sezer, B.; Bilge, G.; Boyaci, I.H. Laser-induced breakdown spectroscopy based protein assay for cereal samples. *J. Agric. Food Chem.* **2016**, *64*, 9459–9463. [CrossRef] [PubMed]
28. Markiewicz-Keszycka, M.; Cama-Moncunill, X.; Casado-Gavalda, M.P.; Dixit, Y.; Cama-Moncunill, R.; Cullen, P.J.; Sullivan, C. Laser-induced breakdown spectroscopy (LIBS) for food analysis: A review. *Trends Food Sci. Technol.* **2017**, *65*, 80–93. [CrossRef]
29. Sezer, B.; Durna, S.; Bilge, G.; Berkkan, A.; Yetisemiyen, A.; Boyaci, I.H. Identification of milk fraud using laser-induced breakdown spectroscopy (LIBS). *Int. Dairy J.* **2018**, *81*, 1–7. [CrossRef]
30. Velioglu, H.M.; Sezer, B.; Bilge, G.; Baytur, S.E.; Boyaci, I.H. Identification of offal adulteration in beef by laser induced breakdown spectroscopy (LIBS). *Meat Sci.* **2018**, *138*, 28–33. [CrossRef]
31. Velásquez-Ferrín, A.; Babos, D.V.; Marina-Montes, C.; Anzano, J. Rapidly growing trends in laser-induced breakdown spectroscopy for food analysis. *Appl. Spectrosc. Rev.* **2021**, *56*, 492–512. [CrossRef]
32. Miziolek, A.W.; Palleschi, V.; Schechter, I. *Laser-Induced Breakdown Spectroscopy (LIBS): Fundamentals and Applications*; Cambridge University Press: Cambridge, UK; New York, NY, USA, 2006; p. 620.
33. Cremers, D.A.; Radziemski, L.J. *Handbook of Laser-Induced Breakdown Spectroscopy*; Wiley: Chichester, UK, 2013.
34. Peng, J.; Xie, W.; Jiang, J.; Zhao, Z.; Zhou, F.; Liu, F. Fast quantification of honey adulteration with laser-induced breakdown spectroscopy and chemometric methods. *Foods* **2020**, *9*, 341. [CrossRef]
35. Muhammed Shameem, K.M.; Dhanada, V.S.; Harikrishnan, S.; George, S.D.; Kartha, V.B.; Santhosh, C.; Unnikrishnan, V.K. Echelle LIBS-Raman system: A versatile tool for mineralogical and archaeological applications. *Talanta* **2020**, *208*, 120482. [CrossRef]
36. Dong, M.; Wei, L.; González, J.J.; Oropeza, D.; Chirinos, J.; Mao, X.; Lu, J.; Russo, R.E. Coal discrimination analysis using tandem laser-induced breakdown spectroscopy and laser ablation inductively coupled plasma time-of-flight mass spectrometry. *Anal. Chem.* **2020**, *92*, 7003–7010. [CrossRef] [PubMed]
37. Abdel-Salam, Z.; Al Sharnoubi, J.; Harith, M.A. Qualitative evaluation of maternal milk and commercial infant formulas via LIBS. *Talanta* **2013**, *115*, 422–426. [CrossRef] [PubMed]
38. Yu, X.; Li, Y.; Gu, X.; Bao, J.; Yang, H.; Sun, L. Laser-induced breakdown spectroscopy application in environmental monitoring of water quality: A review. *Environ. Monit. Assess.* **2014**, *186*, 8969–8980. [CrossRef] [PubMed]
39. Bilge, G.; Boyacı, İ.H.; Eseller, K.E.; Tamer, U.; Çakır, S. Analysis of bakery products by laser-induced breakdown spectroscopy. *Food Chem.* **2015**, *181*, 186–190. [CrossRef]
40. Chen, T.; Huang, L.; Yao, M.; Hu, H.; Wang, C.; Liu, M. Quantitative analysis of chromium in potatoes by laser-induced breakdown spectroscopy coupled with linear multivariate calibration. *Appl. Opt.* **2015**, *54*, 7807–7812. [CrossRef]
41. Nufiqurakhmah, N.; Nasution, A.; Suyanto, H. Laser-induced breakdown spectroscopy (LIBS) for spectral characterization of regular coffee beans and Luwak coffee bean. In Proceedings of the Second International Seminar on Photonics, Optics, and Its Applications (ISPhOA 2016), Bali, Indonesia, 24–25 August 2016; p. 101500M.
42. Wang, J.; Zheng, P.; Liu, H.; Fang, L. Classification of Chinese tea leaves using laser-induced breakdown spectroscopy combined with the discriminant analysis method. *Anal. Methods* **2016**, *8*, 3204–3209. [CrossRef]
43. Mbesse Kongbonga, Y.G.; Ghalila, H.; Onana, M.B.; Ben Lakhdar, Z. Classification of vegetable oils based on their concentration of saturated fatty acids using laser induced breakdown spectroscopy (LIBS). *Food Chem.* **2014**, *147*, 327–331. [CrossRef]

44. Bilge, G.; Sezer, B.; Eseller, K.E.; Berberoğlu, H.; Köksel, H.; Boyacı, İ.H. Determination of Ca addition to the wheat flour by using laser-induced breakdown spectroscopy (LIBS). *Eur. Food Res. Technol.* **2016**, *242*, 1685–1692. [CrossRef]
45. Bilge, G.; Velioglu, H.M.; Sezer, B.; Eseller, K.E.; Boyaci, I.H. Identification of meat species by using laser-induced breakdown spectroscopy. *Meat Sci.* **2016**, *119*, 118–122. [CrossRef]
46. Temiz, H.T.; Sezer, B.; Berkkan, A.; Tamer, U.; Boyaci, I.H. Assessment of laser induced breakdown spectroscopy as a tool for analysis of butter adulteration. *J. Food Compos. Anal.* **2018**, *67*, 48–54. [CrossRef]
47. Markiewicz-Keszycka, M.; Zhao, M.; Cama-Moncunill, X.; El Arnaout, T.; Becker, D.; O'Donnell, C.; Cullen, P.J.; Sullivan, C.; Casado-Gavalda, M.P. Rapid analysis of magnesium in infant formula powder using laser-induced breakdown spectroscopy. *Int. Dairy J.* **2019**, *97*, 57–64. [CrossRef]
48. Nespeca, M.G.; Vieira, A.L.; Júnior, D.S.; Neto, J.A.G.; Ferreira, E.C. Detection and quantification of adulterants in honey by LIBS. *Food Chem.* **2020**, *311*, 125886. [CrossRef]
49. Zar Pasha, A.; Anwer Bukhari, S.; Ali El Enshasy, H.; El Adawi, H.; Al Obaid, S. Compositional analysis and physicochemical evaluation of date palm (*Phoenix dactylifera* L.) mucilage for medicinal purposes. *Saudi J. Biol. Sci.* **2022**, *29*, 774–780. [CrossRef] [PubMed]
50. Moncayo, S.; Rosales, J.D.; Izquierdo-Hornillos, R.; Anzano, J.; Caceres, J.O. Classification of red wine based on its protected designation of origin (PDO) using Laser-induced Breakdown Spectroscopy (LIBS). *Talanta* **2016**, *158*, 185–191. [CrossRef] [PubMed]
51. Liu, F.; Wang, W.; Shen, T.; Peng, J.; Kong, W. Rapid identification of kudzu powder of different origins using laser-induced breakdown spectroscopy. *Sensors* **2019**, *19*, 1453. [CrossRef] [PubMed]
52. Yan, J.; Yang, P.; Hao, Z.; Zhou, R.; Li, X.; Tang, S.; Tang, Y.; Zeng, X.; Lu, Y. Classification accuracy improvement of laser-induced breakdown spectroscopy based on histogram of oriented gradients features of spectral images. *Opt. Express* **2018**, *26*, 28996–29004. [CrossRef]
53. Wadood, S.A.; Nie, J.; Li, C.; Rogers, K.M.; Khan, A.; Khan, W.A.; Qamar, A.; Zhang, Y.; Yuwei, Y. Rice authentication: An overview of different analytical techniques combined with multivariate analysis. *J. Food Compos. Anal.* **2022**, *112*, 104677. [CrossRef]
54. Caceres, J.O.; Moncayo, S.; Rosales, J.D.; de Villena, F.J.M.; Alvira, F.C.; Bilmes, G.M. Application of laser-induced breakdown spectroscopy (LIBS) and neural networks to olive oils analysis. *Appl. Spectrosc.* **2013**, *67*, 1064–1072. [CrossRef]
55. Gazeli, O.; Bellou, E.; Stefas, D.; Couris, S. Laser-based classification of olive oils assisted by machine learning. *Food Chem.* **2020**, *302*, 125329. [CrossRef] [PubMed]
56. Gyftokostas, N.; Stefas, D.; Couris, S. Olive oils classification via laser-induced breakdown spectroscopy. *Appl. Sci.* **2020**, *10*, 3462. [CrossRef]
57. Gondhalekar, C.; Rajwa, B.; Bae, E.; Patsekin, V.; Sturgis, J.; Kim, H.; Doh, I.-J.; Diwakar, P.; Robinson, J.P. Multiplexed detection of lanthanides using laser-induced breakdown spectroscopy: A survey of data analysis techniques. In Proceedings of the SPIE Defense + Commercial Sensing (Sensing for Agriculture and Food Quality and Safety XI), Baltimore, MD, USA, 16–17 April 2019.
58. Gondhalekar, C.; Biela, E.; Rajwa, B.; Bae, E.; Patsekin, V.; Sturgis, J.; Reynolds, C.; Doh, I.-J.; Diwakar, P.; Stanker, L.; et al. Detection of E. coli labeled with metal-conjugated antibodies using lateral-flow assay and laser-induced breakdown spectroscopy. *Anal. Bioanal. Chem.* **2020**, *412*, 1291–1301. [CrossRef]
59. Lu, S.; Shen, S.; Huang, J.; Dong, M.; Lu, J.; Li, W. Feature selection of laser-induced breakdown spectroscopy data for steel aging estimation. *Spectrochim. Acta Part B At. Spectrosc.* **2018**, *150*, 49–58. [CrossRef]
60. Zou, H.; Hastie, T. Regularization and variable selection via the elastic net. *J. R. Stat. Soc. Ser. B (Stat. Methodol.)* **2005**, *67*, 301–320. [CrossRef]
61. Hastie, T.; Tibshirani, R.; Friedman, J. Linear Methods for Regression. In *The Elements of Statistical Learning: Data Mining, Inference, and Prediction*; Springer: New York, NY, USA, 2009; pp. 43–99.
62. Boucher, T.F.; Ozanne, M.V.; Carmosino, M.L.; Dyar, M.D.; Mahadevan, S.; Breves, E.A.; Lepore, K.H.; Clegg, S.M. A study of machine learning regression methods for major elemental analysis of rocks using laser-induced breakdown spectroscopy. *Spectrochim. Acta Part B At. Spectrosc.* **2015**, *107*, 1–10. [CrossRef]
63. Zhang, T.-L.; Wu, S.; Tang, H.-S.; Wang, K.; Duan, Y.-X.; Li, H. Progress of chemometrics in laser-induced breakdown spectroscopy analysis. *Chin. J. Anal. Chem.* **2015**, *43*, 939–948. [CrossRef]
64. Moncayo, S.; Manzoor, S.; Navarro-Villoslada, F.; Caceres, J.O. Evaluation of supervised chemometric methods for sample classification by laser induced breakdown spectroscopy. *Chemom. Intell. Lab. Syst.* **2015**, *146*, 354–364. [CrossRef]
65. Li, L.-N.; Liu, X.-F.; Yang, F.; Xu, W.-M.; Wang, J.-Y.; Shu, R. A review of artificial neural network based chemometrics applied in laser-induced breakdown spectroscopy analysis. *Spectrochim. Acta Part B At. Spectrosc.* **2021**, *180*, 106183. [CrossRef]
66. Wang, Z.; Afgan, M.S.; Gu, W.; Song, Y.; Wang, Y.; Hou, Z.; Song, W.; Li, Z. Recent advances in laser-induced breakdown spectroscopy quantification: From fundamental understanding to data processing. *TrAC Trends Anal. Chem.* **2021**, *143*, 116385. [CrossRef]
67. National Institute for Standards and Technology. NIST Atomic Spectra Database. Available online: https://www.nist.gov/pml/atomic-spectra-database (accessed on 12 December 2022).
68. Sezer, B.; Apaydin, H.; Bilge, G.; Boyaci, I.H. Coffee arabica adulteration: Detection of wheat, corn and chickpea. *Food Chem.* **2018**, *264*, 142–148. [CrossRef]

69. Gamela, R.R.; Costa, V.C.; Sperança, M.A.; Pereira-Filho, E.R. Laser-induced breakdown spectroscopy (LIBS) and wavelength dispersive X-ray fluorescence (WDXRF) data fusion to predict the concentration of K, Mg and P in bean seed samples. *Food Res. Int.* **2020**, *132*, 109037. [CrossRef]
70. Iqbal, J.; Asghar, H.; Shah, S.K.H.; Naeem, M.; Abbasi, S.A.; Ali, R. Elemental analysis of sage (herb) using calibration-free laser-induced breakdown spectroscopy. *Appl. Opt.* **2020**, *59*, 4927–4932. [CrossRef]
71. St-Onge, L.; Kwong, E.; Sabsabi, M.; Vadas, E.B. Rapid analysis of liquid formulations containing sodium chloride using laser-induced breakdown spectroscopy. *J. Pharm. Biomed. Anal.* **2004**, *36*, 277–284. [CrossRef]
72. Díaz Pace, D.M.; D'Angelo, C.A.; Bertuccelli, D.; Bertuccelli, G. Analysis of heavy metals in liquids using Laser Induced Breakdown Spectroscopy by liquid-to-solid matrix conversion. *Spectrochim. Acta Part B At. Spectrosc.* **2006**, *61*, 929–933. [CrossRef]
73. Zhu, D.; Wu, L.; Wang, B.; Chen, J.; Lu, J.; Ni, X. Determination of Ca and Mg in aqueous solution by laser-induced breakdown spectroscopy using absorbent paper substrates. *Appl. Opt.* **2011**, *50*, 5695–5699. [CrossRef] [PubMed]
74. Alamelu, D.; Sarkar, A.; Aggarwal, S.K. Laser-induced breakdown spectroscopy for simultaneous determination of Sm, Eu and Gd in aqueous solution. *Talanta* **2008**, *77*, 256–261. [CrossRef] [PubMed]
75. Sarkar, A.; Alamelu, D.; Aggarwal, S.K. Determination of thorium and uranium in solution by laser-induced breakdown spectrometry. *Appl. Opt.* **2008**, *47*, G58–G64. [CrossRef] [PubMed]
76. Yaroshchyk, P.; Morrison, R.J.S.; Body, D.; Chadwick, B.L. Quantitative determination of wear metals in engine oils using LIBS: The use of paper substrates and a comparison between single- and double-pulse LIBS. *Spectrochim. Acta Part B At. Spectrosc.* **2005**, *60*, 1482–1485. [CrossRef]
77. Keerthi, K.; George, S.D.; Kulkarni, S.D.; Chidangil, S.; Unnikrishnan, V.K. Elemental analysis of liquid samples by laser induced breakdown spectroscopy (LIBS): Challenges and potential experimental strategies. *Opt. Laser Technol.* **2022**, *147*, 107622. [CrossRef]
78. Liu, Y.; Gigant, L.; Baudelet, M.; Richardson, M. Correlation between laser-induced breakdown spectroscopy signal and moisture content. *Spectrochim. Acta Part B At. Spectrosc.* **2012**, *73*, 71–74. [CrossRef]
79. Ayvaz, H.; Sezer, B.; Dogan, M.A.; Bilge, G.; Atan, M.; Boyaci, I.H. Multiparametric analysis of cheese using single spectrum of laser-induced breakdown spectroscopy. *Int. Dairy J.* **2019**, *90*, 72–78. [CrossRef]
80. Clegg, S.M.; Sklute, E.; Dyar, M.D.; Barefield, J.E.; Wiens, R.C. Multivariate analysis of remote laser-induced breakdown spectroscopy spectra using partial least squares, principal component analysis, and related techniques. *Spectrochim. Acta Part B At. Spectrosc.* **2009**, *64*, 79–88. [CrossRef]
81. Geladi, P. Chemometrics in spectroscopy. Part 1. Classical chemometrics. *Spectrochim. Acta Part B At. Spectrosc.* **2003**, *58*, 767–782. [CrossRef]
82. Foegeding, E.A. Food authentication in the 21st century: The power of analytical methods combined with big data analysis. *J. Food Sci.* **2018**, *83*, 1189. [CrossRef]
83. Danezis, G.P.; Tsagkaris, A.S.; Camin, F.; Brusic, V.; Georgiou, C.A. Food authentication: Techniques, trends & emerging approaches. *TrAC Trends Anal. Chem.* **2016**, *85*, 123–132. [CrossRef]
84. Granato, D.; Putnik, P.; Kovačević, D.B.; Santos, J.S.; Calado, V.; Rocha, R.S.; Cruz, A.G.D.; Jarvis, B.; Rodionova, O.Y.; Pomerantsev, A. Trends in chemometrics: Food authentication, microbiology, and effects of processing. *Compr. Rev. Food Sci. Food Saf.* **2018**, *17*, 663–677. [CrossRef] [PubMed]
85. Barron, L.J.R.; Fernández de Labastida, E.; Perea, S.; Chávarri, F.; de Vega, C.; Soledad Vicente, M.a.; Isabel Torres, M.a.; Isabel Nájera, A.; Virto, M.; Santisteban, A.; et al. Seasonal changes in the composition of bulk raw ewe's milk used for Idiazabal cheese manufacture. *Int. Dairy J.* **2001**, *11*, 771–778. [CrossRef]
86. Montel, M.-C.; Buchin, S.; Mallet, A.; Delbes-Paus, C.; Vuitton, D.A.; Desmasures, N.; Berthier, F. Traditional cheeses: Rich and diverse microbiota with associated benefits. *Int. J. Food Microbiol.* **2014**, *177*, 136–154. [CrossRef]
87. Quigley, L.; O'Sullivan, O.; Stanton, C.; Beresford, T.P.; Ross, R.P.; Fitzgerald, G.F.; Cotter, P.D. The complex microbiota of raw milk. *FEMS Microbiol. Rev.* **2013**, *37*, 664–698. [CrossRef] [PubMed]
88. Sánchez-Gamboa, C.; Hicks-Pérez, L.; Gutiérrez-Méndez, N.; Heredia, N.; García, S.; Nevárez-Moorillón, G.V. Seasonal influence on the microbial profile of Chihuahua cheese manufactured from raw milk. *Int. J. Dairy Technol.* **2018**, *71*, 81–89. [CrossRef]
89. Van Hekken, D.L.; Drake, M.A.; Tunick, M.H.; Guerrero, V.M.; Molina-Corral, F.J.; Gardea, A.A. Effect of pasteurization and season on the sensorial and rheological traits of Mexican Chihuahua cheese. *Dairy Sci. Technol.* **2008**, *88*, 525–536. [CrossRef]
90. Dundar, M.; Hirleman, E.D.; Bhunia, A.K.; Robinson, J.P.; Rajwa, B. Learning with a non-exhaustive training dataset: A case study: Detection of bacteria cultures using optical-scattering technology. In Proceedings of the 15th ACM SIGKDD International Conference on Knowledge Discovery and Data Mining, Paris, France, 28 June–1 July 2009; pp. 279–288.
91. Rajwa, B.; Dundar, M.M.; Akova, F.; Bettasso, A.; Patsekin, V.; Dan Hirleman, E.; Bhunia, A.K.; Robinson, J.P. Discovering the unknown: Detection of emerging pathogens using a label-free light-scattering system. *Cytom. Part A* **2010**, *77A*, 1103–1112. [CrossRef] [PubMed]

 *foods*

MDPI

Communication

Evaluating the Use of a Similarity Index (SI) Combined with near Infrared (NIR) Spectroscopy as Method in Meat Species Authenticity

Daniel Cozzolino [1,*], Daniel Bureš [2,3] and Louwrens C. Hoffman [1]

1 Centre for Nutrition and Food Sciences (CNAFS), Queensland Alliance for Agriculture and Food Innovation (QAAFI), The University of Queensland, Brisbane, QLD 4072, Australia
2 Institute of Animal Science, 104 00 Přátelství 815, 104 00 Prague, Czech Republic
3 Department of Food Science, Faculty of Agrobiology, Food and Natural Resources, Czech University of Life Sciences Prague, 165 00 Prague, Czech Republic
* Correspondence: d.cozzolino@uq.edu.au

Abstract: A hand-held near infrared (NIR) spectrophotometer combined with a similarity index (SI) method was evaluated to identify meat samples sourced from exotic and traditional meat species. Fresh meat cuts of lamb (*Ovis aries*), emu (*Dromaius novaehollandiae*), camel (*Camelus dromedarius*), and beef (*Bos taurus*) sourced from a commercial abattoir were used and analyzed using a hand-held NIR spectrophotometer. The NIR spectra of the commercial and exotic meat samples were analyzed using principal component analysis (PCA), linear discriminant analysis (LDA), and a similarity index (SI). The overall accuracy of the LDA models was 87.8%. Generally, the results of this study indicated that SI combined with NIR spectroscopy can distinguish meat samples sourced from different animal species. In future, we can expect that methods such as SI will improve the implementation of NIR spectroscopy in the meat and food industries as this method can be rapid, handy, affordable, and easy to understand for users and customers.

Keywords: exotic species; similarity index; meat; NIR; chemometrics

Citation: Cozzolino, D.; Bureš, D.; Hoffman, L.C. Evaluating the Use of a Similarity Index (SI) Combined with near Infrared (NIR) Spectroscopy as Method in Meat Species Authenticity. *Foods* **2023**, *12*, 182. https://doi.org/10.3390/foods12010182

Academic Editor: Ana M Vivar-Quintana

Received: 16 November 2022
Revised: 22 December 2022
Accepted: 23 December 2022
Published: 1 January 2023

1. Introduction

Red meat represents a significant proportion of the humans' daily diet as it provides nutrients such as protein, vitamins, and minerals, which are essential to maintain a healthy life [1–5]. Different sources of red meats are available and used as a supply of protein such as pork, beef, and lamb as well as wild species in some countries [4–8]. With the growing consumption of red meat and meat products, the consumer is more aware of issues associated with meat safety such as authenticity [3,9–11].

The most recent issues associated with both meat authenticity and fraud involved the replacement of high-value ingredients with not-expensive ones such as horse (e.g., horse meat scandal) [3,12–16]. In other cases, authenticity is associated with the consumption of certain species proscribed by religious reasons (e.g., pork in Muslim countries). The meat industry is also driven by the need to supply the consumer with a consistent high-quality product at an affordable price [4]. Consequently, these issues have increased awareness about authenticity and fraud in the meat industry [3].

Authentication and the recognition of species have been a major threat for the modern meat industry, as it decreases the quality and safety of the meat products [17,18]. Testing of animal meat species is essential for evaluating quality and safeguards the consumer against fraudulent activities [17,18]. This is also of importance to guarantee integrity throughout the supply and value chains. Different analytical methods are available and used for meat identification and authentication, including manual inspection, chromatographic methods (e.g., chromatography mass spectrometry), electrophoretic separation of proteins, molecular-biology-based methods, electronic noses, and vibrational spectroscopy (e.g., near, mid, and Raman spectroscopy) [17–19]. Some of these methods are subjective (e.g., manual inspection), tedious, time consuming, and inconsistent, while other such as molecular biology-based methods (e.g., DNA based techniques, polymerase chain reaction (PCR), real-time PCR, and multiplex PCR), although precise, are slow and expensive [17–19]. Despite these issues associated with the use of traditional methods, vibrational spectroscopy is still considered an emerging technology, which has been proved to be a dynamic and developing method in evaluating and monitoring the authenticity of animal species.

One of the main drawbacks on the utilization of NIR spectroscopy by the food and meat industry is the need of chemometrics to analyze the data collected to make meaningful decisions about the quality and safety of the meat. Chemometrics techniques such as principal component analysis (PCA), discriminant analysis (DA), soft independent modelling of class analogies (SIMCA), and artificial neural networks (ANN) are commonly used to unravel and interpret the spectral properties of the sample, allowing for the classification of samples without the use of direct chemical compositional information [19]. These chemometric techniques have been shown to be able to classify foods, including meat, based on spectral data. However, these advanced chemometrics methods can be difficult to understand and to apply under industrial conditions.

Unlike chemometrics, other qualitative methods, particularly those based on similarity, can be applied to analyze NIR data using "spectral similarity" techniques [20–22]. A simple approach for comparing two spectra is the so-called "similarity index" (SI) method, as described by different authors [20–22]. The SI method has been used and described to identify pure chemicals (e.g., sugar solutions) [20], to compare and authenticate wines [21], as well as to analyze tobacco leaves [22]. The SI method is created using the measurements of the absorbance for every wavelength of the first spectrum, defined as X variable, where the second spectrum is defined as Y variable. The correlation coefficient (r) is used to compute a similarity index which can be used to test for identity between the samples. In this study, NIR spectra are obtained from a meat sample from a given animal species, then a second meat sample from the same or different animal species is taken, and then the two are correlated to confirm or not the authenticity of the meat sample.

This paper details the application of a similarity index (SI) combined with the near infrared (NIR) spectra of meat samples collected using a hand-held spectrophotometer as a rapid, inexpensive tool to authenticate meat samples sourced from traditional and wild meat species.

2. Materials and Methods

Samples of lamb (*Ovis aries*), emu (*Dromaius novaehollandiae*), camel (*Camelus dromedarius*), and beef (*Bos taurus*) were obtained from chilled carcasses after 24 h slaughter and sourced from a commercial slaughterhouse (Queensland, Australia). The fresh meat samples were first cut in small pieces with a knife, and thoroughly hand mixed before being minced. Then, samples were minced using a Tabletop mincer (MEFE 360MC120, 18,000 rpm) fitted with a round mincer plat with 4 mm diameter holes (Mitchell Engineering Food Equipment, Clontorf, Queensland, Australia) which was washed and dried between samples. Four replicates for each species were created (4 animal species $\times$ 4 biological replicates = 16).

The near infrared spectra of the minced meat samples were collected using a hand-held NIR spectrophotometer (Micro-NIR 1700. Viavi, Milpitas, CA, USA) operating in the wavelength range between 950 and 1600 nm (10 nm wavelength resolution). The spectra collection and instrument set up were controlled using the proprietary software provided by the instrument manufacturer (MicroNIR Prov 3.1, Viavi, Milpitas, CA, USA). The spectral data acquisition settings were set at 50 ms integration time and an averaging of 50 scans (MicroNIR Prov 3.1, Viavi, Milpitas, CA, USA). Between samples, a reference spectrum was collected using Spectralon®. The total number of samples used/scanned was 96 (4 animal species $\times$ 4 biological replicates $\times$ 6 scans).

The NIR data were exported into The Unscrambler (version X, CAMO, Oslo, Norway) for data analysis and preprocessing. The NIR spectra was preprocessed using the Savitzky-Golay second derivative (21 smoothing points and second polynomial order) prior to spectra interpretation and chemometric analysis [23]. Principal component analysis (PCA) was used to analyze the data and to evaluate the differences or trends associated with the animal species analyzed. The PCA model was developed and validated using full cross validation (leave one out) [24–26]. Linear discriminant analysis (LDA) was also used to classify meat samples according to the animal species.

In this study, the Similarity Index (SI) method was used to identify and authenticate the meat species analyzed. The SI is specifically targeted to applications whereby only two spectra are being compared (e.g., beef1 vs. beef2). In this index, the r^2 is calculated as the coefficient of determination between the absorbance values from the two spectra at each wavelength across the entire wavelength range. This can be easily determined by use of the correlation function in Excel. The inverse relationship with r^2 means that SI is very sensitive to small changes in r^2, and SI can range in values from 1 (totally different spectra) to infinity ∞ (identical spectra) [20].

$$SI = 1/(1 - r^2)$$

3. Results and Discussion

As the first step, we have interpreted the main features of the NIR spectra of the meat samples analyzed. Figure 1 shows the average of the second derivative NIR spectra of the meat species (e.g., beef, camel, emu, and lamb) analyzed using a hand-held instrument. The second derivative of the NIR spectra of the meat samples showed bands around 985 nm associated with the O-H overtones of water, at 1180 nm (C-H and C=O), at 1205 nm corresponding to a stretching–bending, second overtone of C-H bonds related to lipids [6,9,27]. Additionally, a shoulder around 1350 nm and at 1428 nm O-H stretch first overtone, an O-H combination, and an O-H bend second overtone were noted. These three bands are mainly associated with water content [6,9,27].

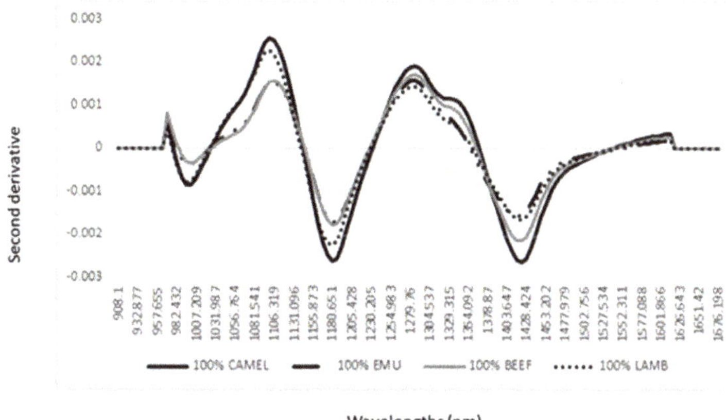

Figure 1. Average second derivative NIR spectra of lamb (*Ovis aries*), emu (*Dromaius novaehollandiae*), camel (*Camelus dromedarius*), and beef (*Bos taurus*) minced samples analyzed using a portable NIR instrument.

A PCA was also performed to observe any trends in the NIR spectra associated with the different meat animal species analyzed. The PCA score plot derived from the analysis of meat samples is shown in Figure 2. The first four principal components (PC) explained 99% of the total variability in the NIR spectra of the meat samples analyzed (PC1 66%, PC2 25%, PC3 5%, and PC4 3%). A separation between meat samples according to the animal species was observed when PC2 vs. PC4 were plotted (Figure 2). The PCA loadings for the first and fourth principal components are reported in Figure 3. The highest loadings in PC2 were observed around 1087 nm (O-H), 1217 nm and 1297 nm (C-H), and at 1428 nm (O-H), while the highest loadings in PC4 were observed at 1050 nm, around 1210 nm (C-H), and 1360 nm (C-H), associated with the presence of lipids (e.g., fatty acid profile) and protein content [6,9,27]. The use of PCA allowed for the identification of differences in the NIR spectra of the meat samples according to the animal species analyzed. These results showed that there is relevant information (e.g., chemical properties) in the NIR spectra that can be used to separate the different animal meat species analyzed.

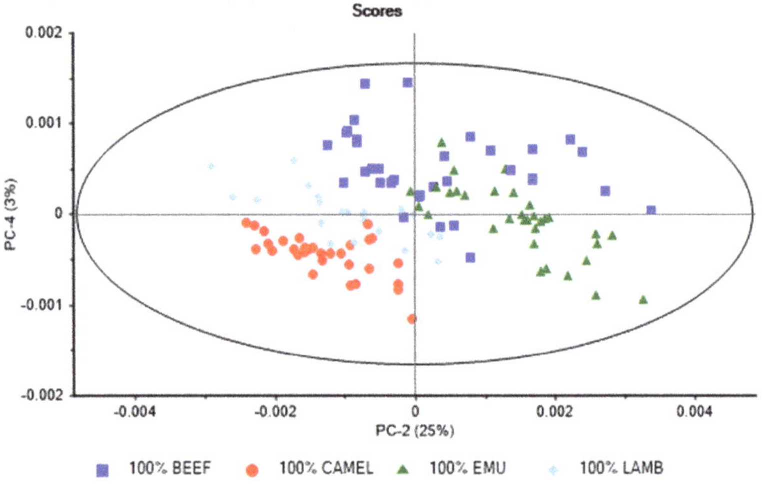

Figure 2. Principal component score plot of NIR spectra of lamb (*Ovis aries*), emu (*Dromaius novaehollandiae*), camel (*Camelus dromedarius*), and beef (*Bos taurus*) minced samples analyzed using a portable NIR instrument.

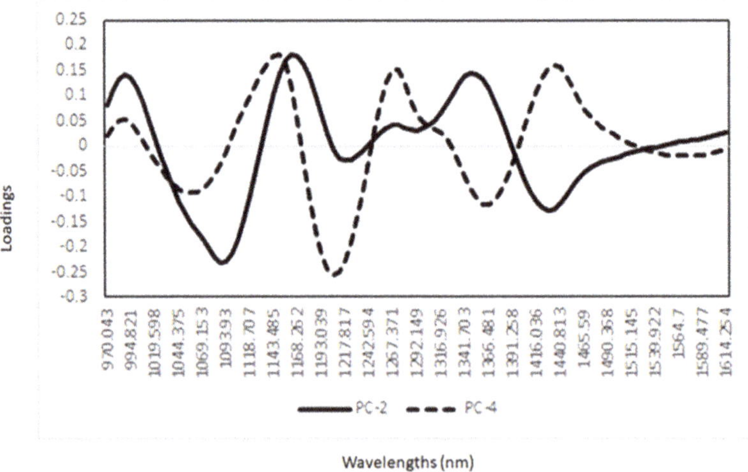

Figure 3. Principal component loadings derived from the analysis of minced meat samples.

In addition to the PCA, the NIR spectra of the meat samples were analyzed using linear discriminant analyses (LDA). The second derivative described in the materials and methods was used as a preprocessing method before LDA. The LDA (using 9 latent variables) confusion matrix obtained from the analysis of the meat samples is shown in Table 1. The overall accuracy of the models was 87.8%. It was observed that 92%, 89%, 86%, and 84% of the camel, emu, beef, and lamb meat samples, respectively, analyzed using NIR spectroscopy were correctly classified.

Table 1. Linear discriminant analysis confusion matrix of meat samples analyzed using near infrared reflectance spectroscopy. In brackets are the percentages of correct classification.

	Camel	Emu	Beef	Lamb
Camel	34 (92%)	0	0	3
Emu	0	33 (89%)	3	0
Beef	4	0	32 (86%)	1
Lamb	2	0	3	31 (84%)

After both PCA and LDA analysis, the similarity index (SI) was calculated. As defined in the previous section, similar samples will have a correlation very close to one (r = 1.0). The SI calculated in this study was considered a more sensitive measure of similarity in comparison with other methods as reported by other authors [20]. In this way, an SI will differ between 1.0 for totally different spectra (e.g., different meat species) and infinity for identical species. In this study, the SI calculated according to previous reports [20,21] was chosen as the indicator of similarity between the same species of meat (e.g., beef1 vs beef 2) [20,21]. The results of the SI for the comparison of the meat samples analyzed using the whole NIR spectra (950 to 1600 nm) are shown in Figure 4, Panel A. The results showed that an SI value > 350 corresponds to a r^2 value > 0.997. This result was considered adequate to either identify the different meat species or the similar ones. Therefore, this value was set as the minimum value for similarity, meaning that the meat samples from the same animal species will have at least a value equal or higher than 350, where meat samples having values below this limit were considered different. It was also observed that the SI successfully matched all identical meat samples corresponding to the same species (e.g., beef vs. beef; lamb vs. lamb, etc.).

Panel A

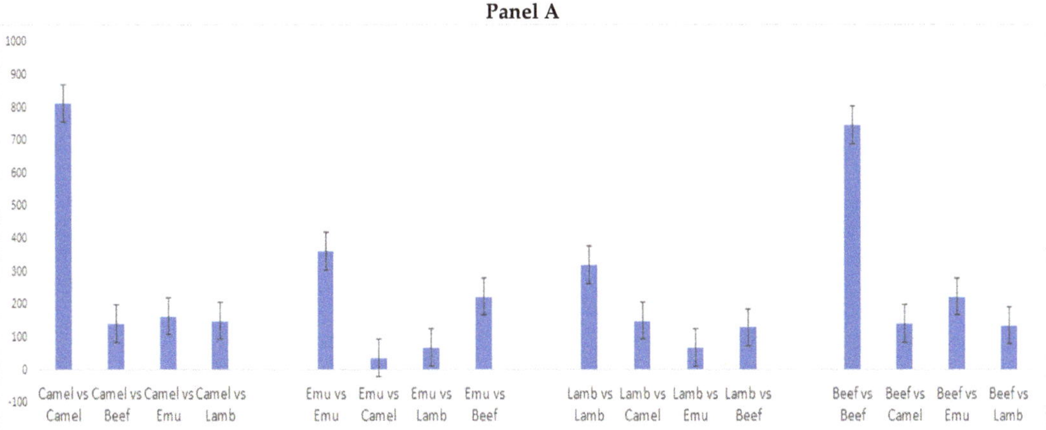

Figure 4. *Cont.*

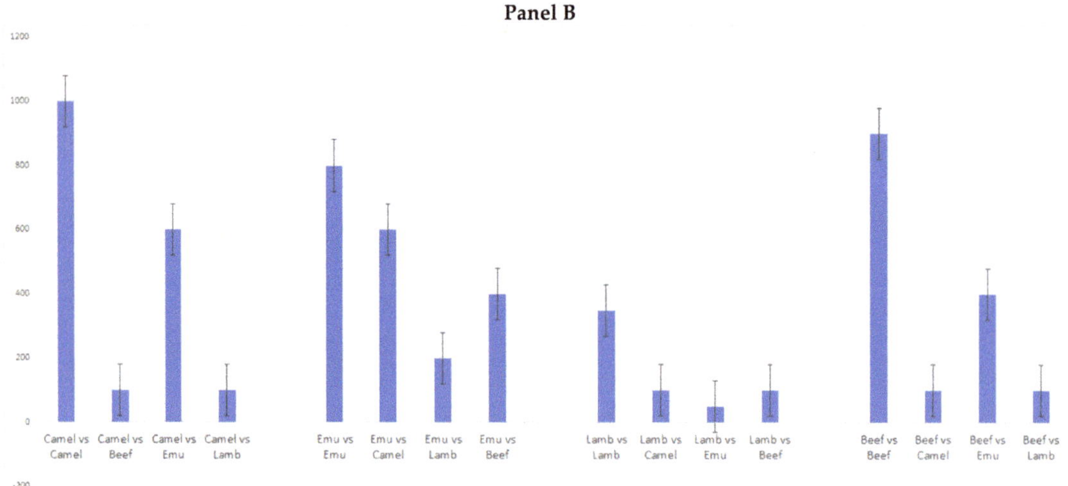

Figure 4. Similarity index calculated for the minced meat samples of lamb (*Ovis aries*), emu (*Dromaius novaehollandiae*), camel (*Camelus dromedarius*), and beef (*Bos taurus*) analyzed using a portable NIR instrument. (**Panel A**): whole range (950 to 1600 nm); (**Panel B**): lipid and protein range (1200 to 1400 nm).

Figure 4, Panel B shows the results for the SI calculated using the NIR range between 1240 and 1400 nm associated with the C-H bonds, related to the range corresponding to lipid and protein contents [6,9,26]. The trends observed were similar to those described in Figure 4, although the threshold has changed to SI > 750 in most of the samples and mixtures analyzed. However, for the meat samples obtained from lamb, the threshold was lower. This can be also due to differences in fat content. Overall, it can be stated that if a sample has a similarity index over 350, the samples can be considered as identical. In instances where the spectra are to a lesser extent easily separated, further research is needed to determine what number is best to be used as the SI.

The utilization of the similarity index method applied to the NIR spectra of meat samples has been proven to be an alternative tool to other classification methods to distinguish samples from the same animal species from different ones (e.g., traditional vs. wild meat). However, research into the overall use of the SI method should also be extended to observe the effects of mixtures, breeds, etc., as it seems the SI is dependent on the data set and experimental conditions. As described above, this simple approach (SI) for comparing two spectra has been described by other authors using different food matrices [20–22]. Overall, the results from this study were comparable with those studies that have analyzed liquid samples such as wine, tobacco leaves, and sugars [20–22]. In addition, the results obtained in this study from the application of the SI are comparable to those obtained using classical chemometrics methods such as LDA (Table 1). The main advantage of using SI over classical chemometric methods (e.g., LDA, PCA) is that this index can be easily understood by the nonexpert where various operators in the industry with a diverse skill base can use this method to trace the origin of the meat. In addition, an SI system must be able to be integrated and operated using readily available equipment (e.g., portable NIR instruments). Ultimately, the use of an SI can be inexpensive and can be implemented on commonly used software such as Excel®.

4. Conclusions

In this study, the use of an SI method to compare the NIR spectra of different animal species was evaluated. The SI method has shown that it can be used as an alternative to other classification methods available such as linear discriminant analysis. Overall, these results indicate that SI combined with NIR spectroscopy can distinguish meat samples sourced from different animal species (e.g., traditional vs. wild meat species). In future, we can expect that methods such as SI will improve the implementation of NIR spectroscopy in the meat and food industries as an authentication tool that is quick, handy, and affordable for customers.

Author Contributions: L.C.H., D.B. and D.C. were involved in the data collection, D.C. was responsible for the data analyses and interpretation. All authors have read and agreed to the published version of the manuscript.

Funding: Internal University of Queensland Institutional funds. The support of the Ministry of Agriculture of the Czech Republic (MZE-RO0718) is also acknowledged.

Institutional Review Board Statement: Not applicable.

Informed Consent Statement: Not applicable.

Data Availability Statement: The data are available from the corresponding author.

Acknowledgments: The authors wish to acknowledge and thank the staff of QAAFI for their help and use of their facilities during the collection of the data.

Conflicts of Interest: The authors declare no conflict of interest.

References

1. Prieto, N.; Roehe, R.; Lavin, P.; Batten, G.; Andres, S. Application of near infrared reflectance spectroscopy to predict meat and meat products quality: A review. *Meat Sci.* **2009**, *83*, 175–186. [CrossRef] [PubMed]
2. Prieto, N.; Pawluczyk, O.; Edward, M.; Dugan, R.; Aalhus, J.L. A Review of the Principles and Applications of Near-Infrared Spectroscopy to Characterize Meat, Fat, and Meat Products. *Appl. Spectrocopy* **2017**, *7*, 1406–1426. [CrossRef] [PubMed]
3. Pieszczek, L.; Czarnik-Matusewicz, H.; Daszykowski, M. Identification of ground meat species using near-infrared spectroscopy and class modeling techniques—Aspects of optimization and validation using a one-class classification model. *Meat Sci.* **2018**, *139*, 15–24. [CrossRef] [PubMed]
4. Chapman, J.; Elbourne, A.; Truong, V.K.; Cozzolino, D. Shining light into meat—A review on the recent advances in in vivo and carcass applications of near infrared spectroscopy. *Int. J. Food Sci. Technol.* **2020**, *55*, 935–941. [CrossRef]
5. Bai, Y.; Liu, H.; Zhang, B.; Zhang, J.; Wu, H.; Zhao, S.; Qie, M.; Guo, J.; Wang, Q.; Zhao, Y. Research Progress on Traceability and Authenticity of Beef. *Food Rev. Int.* **2021**, 1–21. [CrossRef]
6. Mamani-Linares, L.W.; Gallo, C.; Alomar, D. Identification of cattle, llama and horse meat by near infrared reflectance or transflectance spectroscopy. *Meat Sci.* **2012**, *90*, 378–385. [CrossRef] [PubMed]
7. Dumalisile, P.; Manley, M.; Hoffman, L.; Williams, P.J. Discriminating muscle type of selected game species using near infrared (NIR) spectroscopy. *Food Control.* **2020**, *110*, 106981. [CrossRef]
8. Zhang, Y.; Zheng, M.; Zhu, R.; Ma, R. Detection of adulteration in mutton using digital images in time domain combined with deep learning algorithm. *Meat Sci.* **2022**, *192*, 108850. [CrossRef]
9. Cozzolino, D.; Murray, I. Identification of animal meat muscles by visible and near infrared reflectance spectroscopy. *Lebensm.-Wiss. Und-Technol.* **2004**, *37*, 447–452. [CrossRef]
10. Cozzolino, D. Near Infrared Spectroscopy and Food Authenticity. In *Advances in Food Traceability Techniques and Technologies*; Elsevier: Amsterdam, The Netherlands, 2016; pp. 119–136. [CrossRef]
11. López-Maestresalas, A.; Insausti, K.; Jarén, C.; Pérez-Roncal, C.; Urrutia, O.; Beriain, M.J.; Arazuri, S. Detection of minced lamb and beef fraud using NIR spectroscopy. *Food Control* **2019**, *98*, 465–473. [CrossRef]
12. Boyaci, I.H.; Temiz, H.T.; Uysal, R.S.; Velioglu, H.M.; Yadegari, R.J.; Rishkan, M.M. A novel method for discrimination of beef and horsemeat using Raman spectroscopy. *Food Chem.* **2014**, *148*, 37–41. [CrossRef] [PubMed]
13. Kamruzzaman, M.; Makino, Y.; Oshita, S.; Liu, S. Assessment of Visible Near-Infrared Hyperspectral Imaging as a Tool for Detection of Horsemeat Adulteration in Minced Beef. *Food Bioprocess Technol.* **2015**, *8*, 1054–1062. [CrossRef]
14. Qu, J.-H.; Liu, D.; Cheng, J.-H.; Sun, D.-W.; Ma, J.; Pu, H.; Zeng, X.-A. Applications of Near-infrared Spectroscopy in Food Safety Evaluation and Control: A Review of Recent Research Advances. *Crit. Rev. Food Sci. Nutr.* **2015**, *55*, 1939–1954. [CrossRef] [PubMed]
15. Ropodi, A.I.; Pavlidis, D.E.; Mohareb, F.; Panagou, E.Z.; Nychas, G.-J.E. Multispectral image analysis approach to detect adulteration of beef and pork in raw meats. *Food Res. Int.* **2015**, *67*, 12–18. [CrossRef]
16. Rady, A.; Adedeji, A. Assessing different processed meats for adulterants using visible-near-infrared spectroscopy. *Meat Sci.* **2018**, *136*, 59–67. [CrossRef]

17. Cawthorn, D.-M.; Steinman, H.A.; Hoffman, L.C. A high incidence of species substitution and mislabelling detected in meat products sold in South Africa. *Food Control* **2013**, *32*, 440–449. [CrossRef]
18. Edwards, K.; Manley, M.; Hoffman, L.C.; Williams, P.J. Non-Destructive Spectroscopic and Imaging Techniques for the Detection of Processed Meat Fraud. *Foods* **2021**, *10*, 448. [CrossRef]
19. Ballin, R. Lametsch Analytical methods for authentication of fresh vs. thawed meat-A review. *Meat Sci.* **2008**, *80*, 151–158. [CrossRef]
20. Coene, M.P.D.; Grinter, R.; Davies, A.M.C. The use of quadratic regression in qualitative near infrared and visible spectroscopic analysis. *J. Near Infrared Spectrosc.* **1996**, *4*, 153–161. [CrossRef]
21. Bevin, C.; Fergusson, A.; Perry, W.; Janik, L.J.; Cozzolino, D. Development of a Rapid "Fingerprinting" System for Wine Authenticity by Mid-infrared Spectroscopy. *J. Agric. Food Chem.* **2006**, *54*, 9713–9718. [CrossRef]
22. Bi, Y.; Li, S.; Zhang, L.; Li, Y.; He, W.; Tie, J.; Li, F.; Hao, X.; Tian, X.; Tang, L.; et al. Quality evaluation of flue-cured tobacco by near infrared spectroscopy and spectral similarity method. *Spectrochim. Acta Part A Mol. Biomol. Spectrosc.* **2019**, *215*, 398–404. [CrossRef]
23. Saviztky, A.; Golay, M.J.E. Smoothing and differentiation of data by simplified least squares procedures. *Anal. Chem.* **1964**, *36*, 1627–1639. [CrossRef]
24. Bureau, S.; Cozzolino, D.; Clark, C.J. Contributions of Fourier-transform mid infrared (FT-MIR) spectroscopy to the study of fruit and vegetables: A review. *Postharvest Biol. Technol.* **2019**, *148*, 1–14. [CrossRef]
25. Næs, T.; Isaksson, T.; Fearn, T.; Davies, T. *A User-Friendly Guide to Multivariate Calibration and Classification, 2nd ed*; NIR Chichester: Chichester, UK, 2002; Volume 6. [CrossRef]
26. Williams, P.; Dardenne, P.; Flinn, P. Tutorial: Items to be include in a report on a near infrared spectroscopy project. *J. Near Infrared Spectrosc.* **2017**, *25*, 85–90. [CrossRef]
27. Workman, J.; Weyer, L. *Practical Guide to Interpretive Near-Infrared Spectroscopy*; CRC Press Taylor and Francis Group: Boca Raton, FL, USA, 2008.

MDPI

Communication

Characterisation and Identification of Individual Intact Goat Muscle Samples (*Capra* sp.) Using a Portable Near-Infrared Spectrometer and Chemometrics

Louwrens C. Hoffman [1], Prasheek Ingle [1,2], Ankita Hemant Khole [1,2], Shuxin Zhang [1,2], Zhiyin Yang [1,2], Michel Beya [1], Daniel Bureš [3,4] and Daniel Cozzolino [1,*]

1 Queensland Alliance for Agriculture and Food Innovation (QAAFI), Centre for Nutrition and Food Sciences (CNAFS), The University of Queensland, Brisbane, QLD 4072, Australia
2 School of Agriculture and Food Sciences, The University of Queensland, Brisbane, QLD 4072, Australia
3 Institute of Animal Science, Přátelství 815, 104 00 Prague, Czech Republic
4 Department of Food Science, Faculty of Agrobiology, Food and Natural Resources, Czech University of Life Sciences Prague, 165 00 Prague, Czech Republic
* Correspondence: d.cozzolino@uq.edu.au

Abstract: Adulterated, poor-quality, and unsafe foods, including meat, are still major issues for both the food industry and consumers, which have driven efforts to find alternative technologies to detect these challenges. This study evaluated the use of a portable near-infrared (NIR) instrument, combined with chemometrics, to identify and classify individual-intact fresh goat muscle samples. Fresh goat carcasses (n = 35; 19 to 21.7 Kg LW) from different animals (age, breeds, sex) were used and separated into different commercial cuts. Thus, the *longissimus thoracis et lumborum*, *biceps femoris*, *semimembranosus*, *semitendinosus*, *supraspinatus*, and *infraspinatus* muscles were removed and scanned (900–1600 nm) using a portable NIR instrument. Differences in the NIR spectra of the muscles were observed at wavelengths of around 976 nm, 1180 nm, and 1430 nm, associated with water and fat content (e.g., intramuscular fat). The classification of individual muscle samples was achieved by linear discriminant analysis (LDA) with acceptable accuracies (68–94%) using the second-derivative NIR spectra. The results indicated that NIR spectroscopy could be used to identify individual goat muscles.

Keywords: carcass; chemometrics; classification; goat meat; infrared

Citation: Hoffman, L.C.; Ingle, P.; Khole, A.H.; Zhang, S.; Yang, Z.; Beya, M.; Bureš, D.; Cozzolino, D. Characterisation and Identification of Individual Intact Goat Muscle Samples (*Capra* sp.) Using a Portable Near-Infrared Spectrometer and Chemometrics. *Foods* **2022**, *11*, 2894. https://doi.org/10.3390/foods11182894

Academic Editor: David Bongiorno

Received: 4 September 2022
Accepted: 13 September 2022
Published: 18 September 2022

1. Introduction

Meat identification and authentication is one of the applications for which near-infrared (NIR) spectroscopy is considered a valuable tool, as reported by different authors [1–7]. The utilisation of NIR spectroscopy has been reported by different researchers to have great success in identifying and differentiating between different meat species (e.g., beef, pork, lamb, and chicken) as well as authenticating different homogenized meat muscle samples from the same or different animal species [1–8]. The detection of adulterated, unauthentic, poor-quality, and unsafe meats is still a major task for the meat and food industries [9]. The meat industry as well as consumers have driven efforts to introduce innovative and reliable detection techniques that can ensure the authenticity, quality, and safety of both meat and meat products along the supply and value chains [3,5,10,11].

It has been recognised that the so-called classical analytical techniques are expensive, laborious, time-consuming, and not appropriate to the modern challenges facing the food and meat industries. Therefore, the demand to guarantee the authenticity and safety of both meat and meat products has increased the interest in developing rapid analytical techniques in food and meat industries [2–5]. Among these rapid techniques, vibrational spectroscopic techniques, such as NIR, mid-infrared (MIR), and Raman spectroscopies, are useful for the determination of meat quality and authenticity because of their intrinsic characteristics (e.g., rapid, reliable, non-destructive, green, relatively inexpensive) [2–5].

Although NIR spectroscopy has been applied to different commercial and exotic meats (e.g., beef, lamb, pork, chicken, kangaroo, game, etc.) [12–14], not many reports were found that evaluated the use of this technique to analyse goat meat samples. Only one study has been reported that assessed the ability of NIR spectroscopy to characterise and authenticate the composition of goat meat samples [15]. The authors of this study evaluated the use of NIR spectroscopy to estimate protein, moisture, connective tissue, ash, and fat contents in two goat muscles, *Longissimus thoracis* (LT) and *L. lumborum* (LL), with great success (coefficient of determination > 0.70) [15].

Although the focus has been on the adulteration of meat using cheaper alternative species, few studies have evaluated the adulteration of expensive fresh meat cuts with cheaper cuts in the same animal species [16]. Typically, the more expensive cuts in a carcass differ in quality and composition from the inferior cuts or muscles. It is therefore of value to the industry to be able to distinguish between different muscles in a mixture of meat products (e.g., high- vs low-value muscle or commercial cuts), thereby providing proof of provenance and quality; a fillet steak sold as a high-value product due to its inherent quality characteristics is indeed derived from the *Psoas major* muscle and not from some inferior muscle.

Thus, the aim of this study was to evaluate the use of a portable near-infrared (NIR) instrument combined with linear discriminant analysis (LDA) to identify, as well as classify, individual and intact goat muscle samples.

2. Materials and Methods

2.1. Samples

Fresh goat carcasses (n = 35) from different breeds and sexes (male, female), production systems (including commercial farms), and two different experiments were analysed after being slaughtered in a commercial abattoir in Queensland (Australia). The samples were obtained from two different experiments, where in experiment 1, both male and female goat animals were slaughtered, while in experiment 2, only male goats were analysed. The breeds used in these studies were Boer, Boer crosses, and Australian rangeland goats. The goat carcasses were weighed after 24 h (range of 6 to 28 Kg cold carcass weight) and cut in different commercial cuts (e.g., back leg, chump, flap, loin, rack, shoulder), as described by other authors [17]. In this study, the carcasses were weighed, whereafter the muscles in each commercial cut were anatomically dissected. In total, six muscles were dissected and collected for each of the goat carcasses, namely *longissimus thoracis et lumborum* (LTL), *biceps femoris* (BF), *semimembranosus* (SM), *semitendinosus* (ST), *supraspinatus* (SS), and *infraspinatus* (IS). The total number of muscle samples collected and scanned was 210 (35 goats × 6 muscles each).

2.2. Near-Infrared Spectroscopy

The NIR spectra of the individual goat muscle samples were collected using a portable NIR instrument (Micro-NIR 1700. Viavi, Milpitas, CA, USA) operating in the wavelength range of 950–1600 nm (10 nm wavelength resolution). The spectra collection and instrument set-up were controlled using the proprietary software provided by the instrument manufacturer (Viavi Solutions, 2015, Milipitas, CA, USA). The spectral data acquisition settings were set at a 50 ms integration time and an average of 50 scans (MicroNIR Prov 3.1, Viavi, Milpitas, CA, USA). For every 10 samples, a reference spectrum was collected using Spectralon®. Each muscle was scanned in triplicate, and the average of these spectra was used in further chemometric analysis.

2.3. Chemometrics and Data Analysis

The NIR data were exported into The Unscrambler (version X, CAMO, Norway) for data analysis and pre-processing. The NIR spectra were pre-processed using the Savitzky–Golay second derivative (21 smoothing points and second polynomial order) prior to spectra interpretation and chemometric analysis [18]. In this study, principal component analysis (PCA) and linear discriminant analysis (LDA) were used to analyse and classify the muscle samples according to their origin (e.g., type of muscle or breed). The LDA models were developed using the second-derivative NIR spectra and the muscle types as input variables. Models were developed using full cross-validation (leave one out) [19,20]. In addition, the Kennard–Stone approach was used to select samples to be allocated into a calibration and validation set. The ability of the LDA models to classify samples was evaluated using the percentage of correct (%CC) and incorrect (%IC) classifications using the validation set [19,20].

3. Results and Discussion

3.1. Spectra Interpretation

Figure 1 shows the NIR raw spectra of all muscle samples analysed. The raw NIR spectra of the muscles showed three main bands around 976 nm, 1176 nm, and 1428 nm. These bands were associated with third (976 nm) and second (1428 nm) overtones stretching of the O-H bond of water [12,21], while the band around 1176 nm might be associated with the C-H stretching second overtone, associated either with intramuscular fat or lipid content [22–24]. An effect of scatter can be observed in the NIR raw spectra of the muscle samples, mainly due to the presence of water. Therefore, the second derivative was used to improve the interpretation of the NIR spectra of the muscle samples analysed (Figure 2). In addition, the average of the second derivative of the NIR spectra of each of the individual muscle samples analysed is also reported in Figure 3. The NIR absorbances throughout the wavelength range of the individual muscle samples analysed overlapped where main throughs (bands) were observed at 976 nm, 1167 nm, 1341 nm, and 1420 nm. A possible explanation for this overlapping might be related to the similarities in the anatomical location, as well as similar functionality of some of the muscles analysed [14,22]. For example, both ST and SS tended to differentiate from the other muscles around 976 nm (water content) and 1167 nm [12,21]. In addition to the differences between ST and SS, BF tended to differentiate from the other muscles at 1416 nm (water content). A change in the NIR spectra could also be observed around 1200 nm, which is associated with lipids and proteins, in muscles such as ST, SS, and IS. Other authors have also reported that differences between muscles (e.g., in chicken) can be observed in absorbances around 980 nm related to the O-H second overtone (water), at 1202 nm related to the C-H second overtone (lipids), and at 1456 nm related to the O-H first overtone (water) [22–24]. The band around 970 nm is related to the third overtone stretching of an O-H bond associated with water content [12], while the band around 1143 nm corresponds to the second overtone C-H stretching bonds associated with intramuscular fat and lipids [22]. It is known that the proximate chemical composition of meat is influenced by the sex of the animal, where male animals typically have lower fat and higher moisture content than females [14,25]. Considering that muscles from different goat ages and sex groups were utilized in this study, we can infer that some of the differences observed in the NIR spectra can be associated with the intrinsic differences in intramuscular fat, lipids, and moisture content between animals (age and sex) and muscles (anatomical position and functionality). It has also been observed that some of the muscles overlapped around 1392 nm, associated with the second overtone C-H stretching bond that is related to the lipid content of the samples [22]. Within an animal, muscles are known to differ in their chemical composition, including their moisture and intramuscular fat content [25].

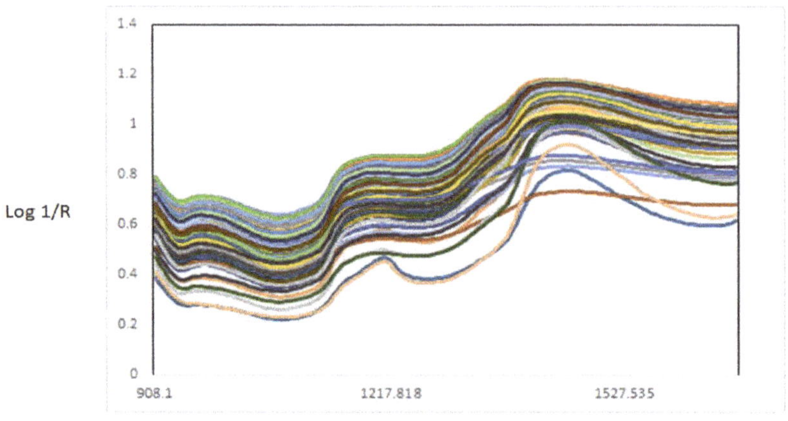

Figure 1. Near-infrared raw spectra of all different intact goat muscle samples analysed.

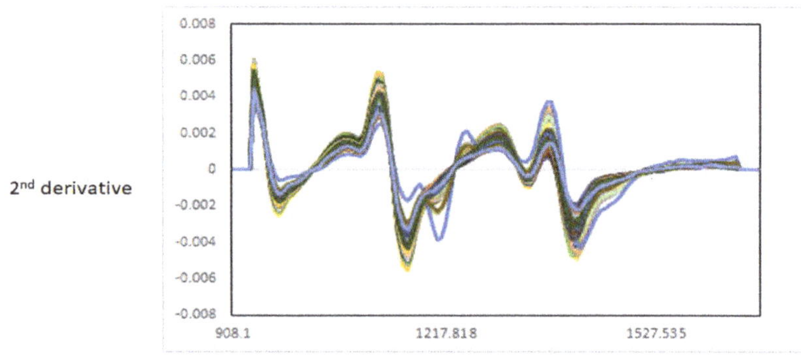

Figure 2. Near-infrared second-derivative spectra of all different intact goat muscle samples analysed.

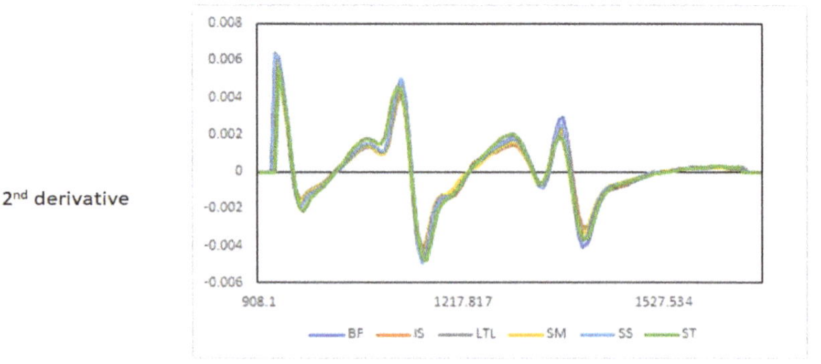

Figure 3. Near-infrared second-derivative average spectra of each of the intact goat muscle samples analysed.

3.2. Principal Component Analysis

Figure 4 shows the PCA score plot and loadings derived from the second-derivative NIR spectra of the intact goat muscle samples analysed. The PCA analysis showed that 94% of the variance in the NIR spectra of the individual muscle samples is explained by the first three principal components (PC1 57%, PC2 32%, and PC3 5%). Although it is not clear from the figures, similar muscle samples tend to cluster together. This trend can also be observed when PC2 vs PC3 are plotted. Muscles such as SM tend to form a tight cluster, while BF and LTL are scattered along the different PCs. Overall, it is difficult to observe a clear separation between the muscle samples when all the samples are analysed together. The highest loadings in PC1 explained the separation between samples and were observed around 976 nm (O-H), 1180 nm (C-H), and 1428 nm (O-H), associated with water content. The highest loadings in both PC2 and PC3 were similar to those observed in PC1, although some shifts in the wavelength were noticeable. The highest loadings in PC3 were observed at 1112 nm, 1180 nm, 1242 nm, and 1397 nm; both bands at 1242 nm and 1397 nm were associated with fat or lipid content [22].

Panel (**A**) PC1 vs. PC2.

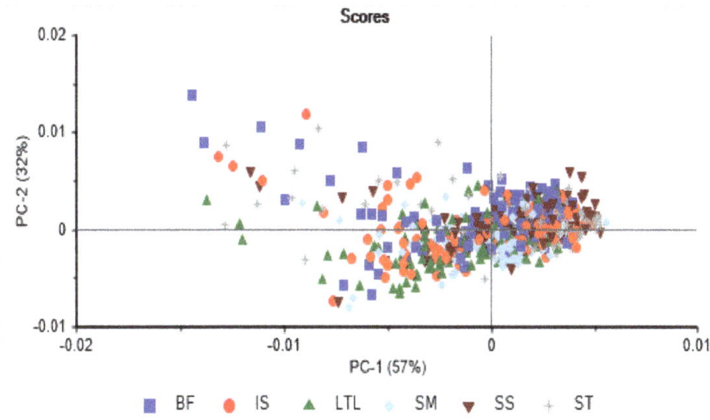

Panel (**B**) PC2 vs. PC3.

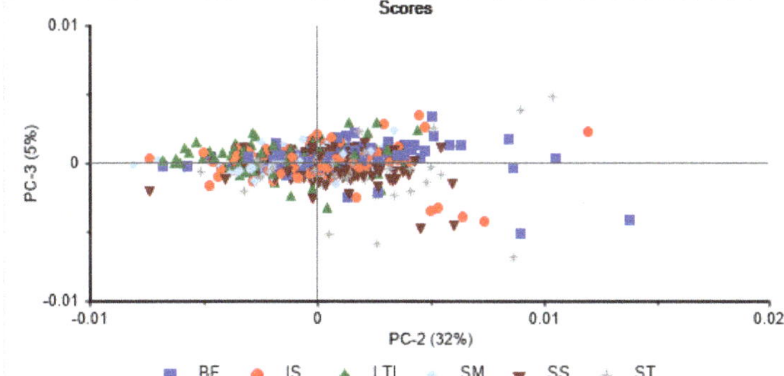

Figure 4. *Cont.*

Panel (**C**) Loadings.

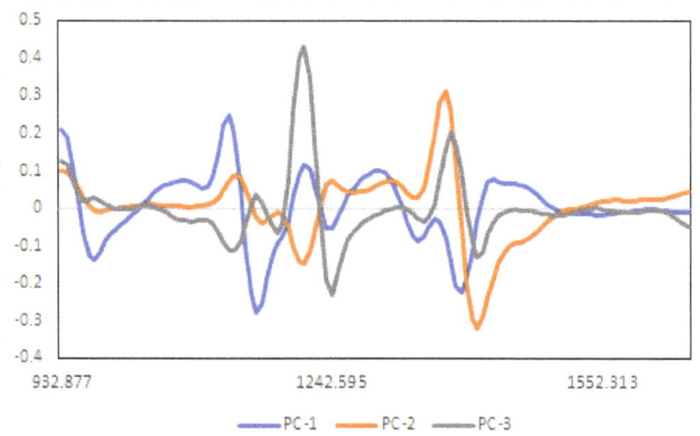

Figure 4. Principal component scores plot (panel (**A**,**B**)) and loadings (panel (**C**)) of intact goat muscles analysed using near-infrared reflectance spectroscopy.

3.3. Classification

The classification results using LDA based on the second-derivative NIR spectra of the individual muscle samples are reported in Table 1. The LDA confusion matrix showed that muscle samples were correctly classified in the range of 63% to 94%, depending on the type of muscle. The poor classification rates were observed for LTL (63%), ST (74%), and SM (71%). For the LTL, 13 samples were misclassified, while 10 and 8 were misclassified for ST and SM, respectively. On the other hand, good to very good classification rates were obtained for BF (82%), SS (94%), and IS (85%), respectively. For BF, six samples were misclassified, while for SS and IS, there were only two and five samples misclassified, respectively. These differences might be attributed to the anatomical and physiological differences among muscles and can also be explained by differences in fibre orientation, muscle chemical composition, physiology, anatomical function, and texture [22,25]. Although the mean second derivative of the NIR spectra appears relatively similar for the different muscle samples analysed, the spectral properties were different, allowing for the discrimination between different muscles.

Table 1. Linear discriminant analysis confusion matrix for the classification of individual goat muscle samples analysed intact by near-infrared reflectance spectroscopy. Results correspond to the validation. In bold is the correct number of samples classified.

	LTL	BF	ST	SM	SS	IS
LTL	**22**	1	0	6	1	5
BF	0	**29**	0	3	2	1
ST	0	1	**26**	1	2	5
SM	1	6	1	**25**	1	1
SS	0	0	0	0	**33**	2
IS	0	0	0	2	3	**30**

LTL: longissimus thoracis et lumborum, BF: biceps femoris, SM: semimembranosus, ST: semitendinosus, SS: supraspinatus, IS: infraspinatus muscles.

We also attempted to discriminate muscles according to genotype (e.g., Boer buck, Boer cross, and Australian rangeland). When all muscle samples were analysed together, a classification rate ranging between 52 and 58% was achieved. Thus, comparisons between Boer buck and Australian rangeland, Boer cross, and Australian rangeland, as well as Boer cross and Boer buck, were made separately. Muscle samples were classified correctly with an 80% rate when Boer buck and Australian rangeland were compared. For the other two groups, although an improvement in the classification rate (correct classification around 70%) was achieved, the muscles belonging to the Boar cross were not correctly classified. This might be explained by the fact that Boer buck and cross goats are more genetically similar compared with the Australian rangeland animals. The results of this study indicated that NIR spectroscopy was able to identify the origin of the muscles using intact samples (thus, there is no need for homogenization). These results indicate that NIR use can also be extended to other species and muscles as a high-throughput tool to identify the origin of the meat.

4. Conclusions

This study reported the use of a portable NIR spectrometer combined with chemometrics to characterise and identify different goat muscle samples. Differences in the NIR spectra of the muscles were observed around 970 nm, 1242 nm, 1397 nm, and 1428 nm associated with water and fat content (e.g., IMF). The classification of individual muscle samples showed that samples could be classified with accuracies ranging from 68% to 94% using the second-derivative NIR spectra. Muscles that are in the same anatomical location, such as the IS and SS, were correctly classified by NIR spectroscopy. Overall, the results of this study indicated that NIR spectroscopy could be used to characterise and identify different intact goat muscle samples. In future, we can expect an improvement in the NIR models by incorporating samples from other commercial and production conditions, as well as different genetics. The findings of this research might be extended to other species and types of muscles produced and sold within a commercial facility with the several advantages NIR provides, such as the low cost and the fact that this technique it is non-destructive.

Author Contributions: L.C.H., P.I., A.H.K., S.Z., Z.Y., M.B., D.B. and D.C.: data collection and analysis; L.C.H., D.B. and D.C.: draft preparation and editing of the manuscript; L.C.H. and D.C.: supervision. L.C.H.: project administration. All authors have read and agreed to the published version of the manuscript.

Funding: The work was partly funded by UQ internal funds. BD received funds from the Ministry of Agriculture, MZE in Czech Ministerstvo Zemědělství ČR, grant number MZE-RO0718.

Institutional Review Board Statement: Not applicable.

Informed Consent Statement: Not applicable.

Data Availability Statement: Not applicable.

Conflicts of Interest: The authors declare no conflict of interest.

References

1. Abbas, O.; Zadravec, M.; Baeten, V.; Mikuš, T.; Lešić, T.; Vulić, A.; Prpić, J.; Jemeršić, L.; Pleadin, J. Analytical methods used for the authentication of food of animal origin. *Food Chem.* **2018**, *246*, 6–17. [CrossRef] [PubMed]
2. Guy, F.; Prache, S.; Thomas, A.; Bauchart, D.; Andueza, D. Prediction of lamb meat fatty acid composition using near infrared reflectance spectroscopy (NIRS). *Food Chem.* **2011**, *127*, 1280–1286. [CrossRef] [PubMed]
3. Kademi, H.I.; Ulusoy, B.H.; Hecer, C. Applications of miniaturized and portable near infrared spectroscopy (NIRS) for inspection and control of meat and meat products. *Food Rev. Int.* **2019**, *35*, 201–220. [CrossRef]
4. Prieto, N.; Roehe, R.; Lavín, P.; Batten, G.; Andrés, S. Application of near infrared reflectance spectroscopy to predict meat and meat products quality: A review. *Meat Sci.* **2009**, *83*, 175–186. [CrossRef] [PubMed]
5. Rady, A.; Adedeji, A. Assessing different processed meats for adulterants using visible-near-infrared spectroscopy. *Meat Sci.* **2018**, *136*, 59–67. [CrossRef] [PubMed]
6. Viljoen, M.; Hoffman, L.C.; Brand, T.S. Prediction of the chemical composition of mutton with near infrared reflectance spectroscopy. *Small Rumin. Res.* **2007**, *69*, 88–94. [CrossRef]
7. Weeranantanaphan, J.; Downey, G.; Allen, P.; Sun, D.W. A review of near infrared spectroscopy in muscle food analysis: 2005–2010. *J. Near Infrared Spectrosc.* **2011**, *19*, 61–104. [CrossRef]

8. Damez, J.L.; Clerjon, S. Quantifying and predicting meat and meat products quality attributes using electromagnetic waves: An overview. *Meat Sci.* **2013**, *95*, 879–896. [CrossRef]
9. Kamruzzaman, M.; Elmasry, G.; Sun, D.W.; Allen, P. Application of NIR hyperspectral imaging for discrimination of lamb muscles. *J. Food Eng.* **2011**, *104*, 332–340. [CrossRef]
10. McVey, C.; Elliot, C.T.; Cannavan, A.; Kelly, S.D.; Petchkongkaew, A.; Simon, A.; Haughey, S.A. Portable spectroscopy for high throughput food authenticity screening: Advancements in technology and integration into digital traceability systems. *Trends Food Sci. Technol.* **2021**, *118*, 777–790. [CrossRef]
11. Cawthorn, D.-M.; Steinman, H.A.; Hoffman, L.C. A high incidence of species substitution and mislabelling detected in meat products sold in South Africa. *Food Control* **2013**, *32*, 440–449. [CrossRef]
12. Barbin, D.; Elmasry, G.; Sun, D.W.; Allen, P. Near-infrared hyperspectral imaging for grading and classification of pork. *Meat Sci.* **2012**, *90*, 259–268. [CrossRef] [PubMed]
13. Dixit, Y.; Casado-Gavalda, M.P.; Cama-Moncunill, R.; Cullen, P.J.; Sullivan, C. Challenges in Model Development for Meat Composition Using Multipoint NIR Spectroscopy from At-Line to In-Line Monitoring. *J. Food Sci.* **2017**, *82*, 1557–1562. [CrossRef] [PubMed]
14. Dumalisile, P.; Manley, M.; Hoffman, L.; Williams, P.J. Discriminating muscle type of selected game species using near infrared (NIR) spectroscopy. *Food Control* **2020**, *110*, 106981. [CrossRef]
15. Teixeira, A.; Oliveira, A.; Paulos, K.; Leite, A.; Marcia, A.; Amorim, A.; Pereira, E.; Silva, S.; Rodrigues, S. An approach to predict chemical composition of goat Longissimus thoracis et lumborum muscle by Near Infrared Reflectance spectroscopy. *Small Rumin. Res.* **2015**, *126*, 40–43. [CrossRef]
16. An, J.; Li, Y.; Zhang, C.; Zhang, D. Rapid non-destructive prediction of multiple quality attributes for different commercial meat cut types using optical system. *Food Sci. Anim. Resour.* **2022**, *42*, 655–6571. [CrossRef]
17. Dieters, L.S.E.; Meale, S.J.; Quigley, S.P.; Hoffman, L.C. Meat quality characteristics of lot-fed Australian Rangeland goats are unaffected by live weight at slaughter. *Meat Sci.* **2021**, *175*, 108437. [CrossRef]
18. Savitzky, A.; Golay, M.J.E. Smoothing and differentiation of data by simplified least squares procedures. *Anal. Chem.* **1964**, *36*, 1627–1639. [CrossRef]
19. Bureau, S.; Cozzolino, D.; Clark, C.J. Contributions of Fourier-transform mid infrared (FT-MIR) spectroscopy to the study of fruit and vegetables: A review. *Postharvest Biol. Technol.* **2019**, *148*, 1–14. [CrossRef]
20. Oliveri, P.; Downey, G. Multivariate class modelling for the verification of food authenticity claims. *TRAC Trends Anal. Chem.* **2012**, *35*, 74–86. [CrossRef]
21. Elmasry, G.; Iqbal, A.; Sun, D.W.; Allen, P.; Ward, P. Quality classification of cooked, sliced Turkey hams using NIR hyperspectral imaging system. *J. Food Eng.* **2011**, *103*, 333–344. [CrossRef]
22. Cozzolino, D.; Murray, I. Identification of animal meat muscles by visible and near infrared reflectance spectroscopy. *LWT-Food Sci. Technol.* **2004**, *37*, 447–452. [CrossRef]
23. Ding, H.; Wang, G.; Lei, W.; Wang, R.; Huang, L.; Xia, Q.; Wu, J. Non-invasive quantitative assessment of oxidative metabolism in quadriceps muscles by near infrared spectroscopy. *Br. J. Sports Med.* **2001**, *35*, 441–444. [CrossRef] [PubMed]
24. Workman, J.; Weer, L. *Practical Guide to Interpretive Near-Infrared Spectroscopy*; CRC Press Taylor and Francis Group: Boca Raton, FL, USA, 2008.
25. Van Wyk, G.L.; Hoffman, L.C.; Strydom, P.E.; Frylinck, L. Differences in meat quality of six muscles obtained from southern African large frame indigenous Veld goat and Boer goat wethers and bucks. *Animals* **2022**, *12*, 382. [CrossRef]

 foods

MDPI

Article

Assessment of the Authenticity of Whisky Samples Based on the Multi-Elemental and Multivariate Analysis

Magdalena Gajek [1,*], Aleksandra Pawlaczyk [1], Elżbieta Maćkiewicz [1], Jadwiga Albińska [1], Piotr Wysocki [1], Krzysztof Jóźwik [2] and Małgorzata Iwona Szynkowska-Jóźwik [1]

1 Faculty of Chemistry, Institute of General and Ecological Chemistry, Lodz University of Technology, Zeromskiego 116, 90-924 Lodz, Poland
2 Faculty of Mechanical Engineering, Institute of Turbomachinery, Lodz University of Technology, Wolczanska 219/223, 90-924 Lodz, Poland
* Correspondence: magdalena.gajek@edu.p.lodz.pl; Tel.: +48-42-631-30-95

Abstract: Two hundred and five samples of whisky, including 170 authentic and 35 fake products, were analyzed in terms of their elemental profiles in order to distinguish them according to the parameter of their authenticity. The study of 31 elements (Ag, Al, B, Ba, Be, Bi, Cd, Co, Cr, Cu, Li, Mn, Mo, Ni, Pb, Sb, Sn, Sr, Te, Tl, U, V, Ca, Fe, K, Mg, P, S, Ti and Zn) was performed using the Inductively Coupled Plasma Mass Spectrometry (ICP-MS), Inductively Coupled Plasma Optical Emission Spectrometry (ICP-OES) and Cold Vapor-Atomic Absorption (CVAAS) techniques. Additionally, the pH values of all samples were determined by pH-meter, and their isotopic ratios of $^{88}Sr/^{86}Sr$, $^{84}Sr/^{86}Sr$, $^{87}Sr/^{86}Sr$ and $^{63}Cu/^{65}Cu$ were assessed, based on the number of counts by ICP-MS. As a result of conducted research, elements, such as Mn, K, P and S, were identified as markers of whisky adulteration related to the age of alcohol. The concentrations of manganese, potassium and phosphorus were significantly lower in the fake samples (which were not aged, or the aging period was much shorter than legally required), compared to the original samples (in all cases subjected to the aging process). The observed differences were related to the migration of these elements from wooden barrels to the alcohol contained in them. On the other hand, the sulfur concentration in the processed samples was much higher in the counterfeit samples than in the authentic ones. The total sulfur content, such as that of alkyl sulfides, decreases in alcohol with aging in the barrels. Furthermore, counterfeit samples can be of variable origin and composition, so they cannot be characterized as one group with identical or comparable features. Repeatedly, the element of randomness dominates in the production of these kinds of alcohols. However, as indicated in this work, the extensive elemental analysis supported by statistical tools can be helpful, especially in the context of detecting age-related adulteration of whisky. The results presented in this paper are the final part of a comprehensive study on the influence of selected factors on the elemental composition of whisky.

Keywords: authentication; adulteration; fake; whisky; elemental analysis; ICP-MS; ICP-OES; CVAAS; spirits; principal component analysis; alcohol aging; isotope ratios

Citation: Gajek, M.; Pawlaczyk, A.; Maćkiewicz, E.; Albińska, J.; Wysocki, P.; Jóźwik, K.; Szynkowska-Jóźwik, M.I. Assessment of the Authenticity of Whisky Samples Based on the Multi-Elemental and Multivariate Analysis. *Foods* **2022**, *11*, 2810. https://doi.org/10.3390/foods11182810

Academic Editor: Daniel Cozzolino

Received: 4 August 2022
Accepted: 29 August 2022
Published: 12 September 2022

1. Introduction

Extremely fast development of trade and international exchange of products and food mobility brought an unprecedented variety of food products to consumers. However, nowadays, consumer awareness regarding the quality and authenticity of the food they buy and consume was raised significantly. Moreover, a study conducted over a decade ago indicated that as many as 82% of the customers considered geographical origin as a quality indicator before purchasing food products [1]. Literature reports clearly suggest that numerous cases of food adulteration have been reported, including the use of substances that pose a threat to the health and life of consumers. Examples of such activities can be given as follows: mixing melamine and wheat gluten to increase the protein content [2], contamination of paprika powder with lead oxide [3], addition of red lead (Pb_3O_4) to cayenne pepper to achieve a vibrant color [4]. In turn, honeys are often adulterated to increase their shelf-life and nutritional value, by adding glucose–fructose syrups, corn syrups, invert sugar syrups or by admixing with imported honeys of poorer quality [5,6]. Thus, food authenticity is an important matter in the case of quality control and assurance of food safety. The authentication of food concerns many aspects, including misleading about origin, mislabeling and adulteration, which is defined as a process by which the quality or the nature of a given product is reduced due to the addition of a foreign or an inferior substance and removing a vital element [7,8].

The need for precise and valid analytical techniques for food investigations is increasing because of the continuously rising food deception around the world [9–11]. Fortunately, a range of potential analytical techniques for the authenticity termination and traceability of food products is extensive. Among them, the following methods can be distinguished: spectroscopic techniques [12–15] (including those based on isotopic ratios [16,17]), separation techniques [6,18], neutron and proton-based nuclear techniques [19], as well as advanced DNA-based techniques [10,20]. Elemental analysis has long been used in research connected with food authenticity, including discrimination of geographical origin [7], organic versus conventional cultivation [21] or free range to compare with conventionally farmed products [22]. Numerous literature reports indicate that elemental fingerprinting also proved its usefulness for the differentiation of origin of wine [15,23], olive oil [24], honey [6,25], coffee [26], tea [27], cheese [28], vegetables and fruits [29] and also spices and food additives [30]. Food products consist of numerous compounds, including carbohydrates, peptides, lipids, fatty acids, amino acids, organic acids, nucleic acids and other small molecules (aromas, dyes, preservatives and other exogenous compounds) [31]. Due to the complexity of the ingredients in the food, using chromatographic methods it makes possible to obtain unique molecular fingerprints, which has a huge potential in differentiation during the authentication process [30]. Separation techniques were used for food authentication and geographic identification of the following: apple juice [32], kiwifruit juices [33], wine [34], honey [6], saffron [35], tomatoes [36], ginger [37], whisky [38–42] and fruit spirits [43]. Moreover, the isotopic ratios were successfully used in food authentication because stable isotope ratios are dependent on the climatic and soil conditions, as well as geographical origin of food ingredients [30]. The isotope ratios mostly investigated in food authentication are $^{2}H/^{1}H$, $^{13}C/^{12}C$, $^{15}N/^{14}N$, $^{18}O/^{16}O$, $^{34}S/^{32}S$, $^{84}Sr/^{86}Sr$, $^{87}Sr/^{86}Sr$, $^{88}Sr/^{86}Sr$ $^{206}Pb/^{204}Pb$, $^{207}Pb/^{204}Pb$ and $^{208}Pb/^{204}Pb$ [44,45]. Literature reports indicate that techniques based on the measurement of isotope ratios are most often used for authentication of cheeses [46], sweet cherries [47], lentils [48] bell pepper [49], wheat [50], wine [51,52] and vodka [53].

Due to the great popularity and high price, premium whisky is one of the most frequently counterfeited alcoholic beverages. The process of counterfeiting whisky usually involves blending a cheaper version of whisky belonging to the same category as the genuine brand, mixing a cheap local alcohol with the original brand of whisky or using a cheap local alcohol with added flavorings and coloring as a genuine product [54]. Another possibility of counterfeits in the case of whisky is the use of a different type of barrel, as well as a much shorter aging period compared to the manufacturer's declarations. The most important quality characteristics, particularly in the case of premium brands, are the maturation period and the history of the casks in which whisky was matured. Thus, during the authentication process of whisky, a number of facts have to be taken into consideration. The water, the cereals, the use of peat smoke during grain malting and the equipment applied in the distillation process will have an influence, to a greater or lesser extent, on the final product. During the aging of the raw distillate in the barrel, significant changes take place in the chemical composition of the alcohol, which results in the "softening" of the product [42,55]. As previously noted, the analytical techniques most commonly used to authenticate and identify the geographical origin of whisky are chromatographic methods [38–42]. They allow finding characteristic compounds and determine aroma profiles, which can then be used to define the quality and authenticity of the tested whisky [56]. Especially the analysis of esters, which have the greatest impact on the aroma of the alcohol, enables an assessment of the aging process and, as a result, the verification of the authenticity of the age of whisky [57,58].

Taking into account the number of scientific studies dealing with the authentication and identification of the origin of food products, most of articles refer to wines; then fruit, vegetables and cereals; and, finally, meats, oils and fats. The available scientific data show that less than 10% of all publications devoted to food authentication concern the analysis of beverages (including spirit, beers, soft drinks and mineral waters) [30]. To the authors' knowledge, very few papers on metal analysis in whisky are available [59–62]. However, the use of the elemental profile to establish authenticity and provenance is extremely rare in the literature [60]. In the first part of the scientific study (The Elemental Fingerprints of Different Types of Whisky as Determined by ICP-OES and ICP-MS Techniques in Relation to Their Type, Age, and Origin [61]), the extensive elemental characterization of whisky samples was performed, including distinguishing alcohol samples based on their origin, type and age using statistical analysis and chemometric tests. The authors in this paper have not discussed the issues related to the authenticity of products or its possible identification.

The main purpose of this work was to assess the authenticity parameter based on an extensive elemental analysis supported by appropriate statistical and chemometric tests. It should be emphasized that in this study wide range of measurements were carried out with the use of 3 analytical techniques (ICP-MS, ICP-OES and CV-AAS) to determine the concentrations of 31 elements in 205 whisky samples (170 authentic and 35 fake samples). Additionally, the pH value was measured for each of the analyzed alcohol samples, and the collected semi-quantitative data were used to determine the isotope ratios.

2. Materials and Methods

2.1. Samples

In this study, a total of 205 whisky samples were analyzed, including 170 samples of original products, which were discussed in the first part of the publication (The Elemental Fingerprints of Different Types of Whisky as Determined by ICP-OES and ICP-MS Techniques in Relation to Their Type, Age, and Origin [63]), as well as 35 samples of unidentified identity, called fake products, which were used as a reference group for the authenticity studies. Among the 35 samples, 9 different sources of their origin can be distinguished. The source of origin is understood to mean the producer or the place where the product was manufactured. These alcohols were distributed on various scales as analog of whisky products. To the authors' knowledge, fake alcohols were not matured in wooden barrels or this stage was significantly reduced. However, the counterfeits whisky products were from sources that remain anonymous. The analysis was performed using the ICP-MS, ICP-OES and CVAAS techniques.

The information about whisky products categories was coded, and the manufacturers' names are not given in this paper. Basic characteristics of the tested samples are included in Table 1.

Table 1. Characteristics of the tested set of samples.

n	Authentic	Fake								
		35								
	170	Number of Samples from a Given Source								
		S1–7	S2–9	S3–6	S4–2	S5–3	S6–4	S7–2	S8–1	S9–1
Total	205									

S1–S9 code of source of origin (e.g., S1—source no 1).

2.2. Samples Preparation and Equipment

- ICP-OES, ICP-MS and CV-AAS

The sample preparation procedures and the measurement conditions are described in detail in the publication Elemental Fingerprint of Different Types of Whisky Determined by ICP-OES and ICP-MS techniques in Relation to Their Type, Age and Origin [61] and in our preliminary study (Multielemental Analysis of Various Kinds of Whisky [63]). Moreover, all validation procedures were analogous to those described in the first part of the paper.

- pH-Metr

Basic 20⁺ pH-meter (CARISON INSTRUMENTS S.A., Barcelona, Spain) was used to measure the pH values of the tested whisky samples. The pH-meter consists of a magnetic stirrer with automatic temperature stabilization and a combined electrode with glass and a silver chloride electrode placed in one holder. Before the measurement, the necessary calibration process was performed using buffers at pH 4.01, 7.00 and 9.21 (HACH Company, Düsseldorf, Germany). Measurements were carried out during a three-day analytical cycle. Three replicates were performed for each sample, and the average result was taken as the final result. After analyzing 20 samples, calibration was repeated.

2.3. Data Analysis

The STATISTICA 12.5 (New York, NY, USA) software was employed for raw data processing. The first step was to check the normality of the distribution of the studied variables. In this order, Kołmogorow–Smirnow tests were applied. On the basis of the tests, the hypothesis of normal distribution was rejected for all studied elements and isotope ratios, as well as pH-value (for the significance level $\alpha = 0.05$). Then, the existence of statistically significant differences was checked. For this purpose, the Kruskal–Wallis non-parametric test was used. In the final phase, data were investigated by multivariate chemometric analysis. To increase the interpretability of the results, principal component analysis (PCA) was applied.

3. Results and Discussion

3.1. Level of Metals in Analyzed Whisky Samples

In this study, the concentration of 31 elements in 205 whisky samples and products of unknown identity was determined. A total of 170 samples are authentic products, the concentrations of which were listed in the first part. The remaining 35 items are false objects and the obtained results for this group regarding their elemental profile were given in this paper. The ICP-MS technique was used to determine the concentration of the following elements: Ag, Al, B, Ba, Be, Bi, Cd, Co, Cr, Cu, Li, Mn, Mo, Ni, Pb, Sb, Sn, Sr, Te, Tl, U and V, while elements, such as Ca, Fe, K, Mg, P, S, Ti and Zn, were measured with the ICP-OES technique. The CVAAS technique was used to determine the total mercury content.

In terms of 35 samples of counterfeit products, some of the obtained results were below the quantification limits. The Hg concentration was below the limit of quantification in each case. Te was not determined in 31 samples. Ag was not determined in 19 samples, P in 15 and Fe in 13. Sb and Bi were not detected in 12 samples, while Cd and Ti in 9 samples. Zn was not found in six samples; Mo and Tl in four; and Al, V, Sn and Pb in three samples. U was not identified in two independent samples, while Li, Be and B were not quantified in one sample.

In the first part of the publication, the basic statistical parameters of authentic products (170 samples) were summarized. Therefore, in Table 2 the same type of the information was given, such as the mean, median, minimum and maximum, but for the group of counterfeit products (35 samples). In each case, due to the rejection of the hypothesis of normal distribution, in order to assess statistically significant differences between the groups under consideration, the non-parametric Kruskal–Wallis test was applied.

Table 2. Basic statistics for determined elements for all counterfeit samples (n = 35) [μg/L].

Element	n	Mean	Median	Min	Max	Element	n	Mean	Median	Min	Max
Ag		1.280	<LOQ	<LOQ	8.600	Sb		0.540	0.300	<LOQ	3.000
Al		168.4	163.3	<LOQ	470.7	Sn		13.89	9.810	<LOQ	34.70
B		3794	2397	<LOQ	19.02	Sr		133.0	53.72	14.146	765.1
Ba		199.6	189.0	117.3	378.0	Te		0.060	<LOQ	<LOQ	1.100
Be		0.130	0.110	<LOQ	0.500	Tl		0.210	0.030	<LOQ	2.100
Bi		3.220	0.600	<LOQ	25.80	U		0.360	0.190	<LOQ	3.100
Cd		6.110	0.760	<LOQ	65.90	V		1.680	0.910	<LOQ	10.40
Co	35	9.920	5.260	1.409	42.20	Ca	35	35.73	22.91	1994	271.1
Cr		182.5	112.3	54.57	770.3	Fe		174.7	29.98	<LOQ	2735
Cu		2383	56.86	1.922	33.21	K		97.09	10.88	<LOQ	670.6
Li		67.12	19.25	<LOQ	825.4	Mg		5370	1577	465.4	33.07
Mn		76.75	51.39	2.377	438.7	P		7352	74.29	<LOQ	56.79
Mo		11.07	1.590	<LOQ	108.4	S		20.89	14.68	197.6	231.7
Ni		62.71	39.86	2.418	411.0	Ti		43.49	25.35	<LOQ	316.8
Pb		12.84	11.21	<LOQ	35.60	Zn		2987	274.8	<LOQ	39.82

The average contents of median values for the elements in the alcohol samples of unidentified origin decreased in the following order: Ca > K > S > B > Mg > Zn > Ba > Al > Cr > Sr > P > Fe > Cu > Ni > Mn > Ti > Li > Sn > Pb > Co > Mo > V > Cd > Bi > Sb > U > Be > Tl > Ag > Te > Hg. The order of elements for authentic samples was similar with the general trend from macro to micro elements. However, it should be noted that in the case of original products, elements, such as P and Cu, are listed higher in this order, while S lower than the presented order for non-original samples.

The authors of this paper referred to the internal national standards that define the maximum permissible content of selected metals (Cd, Pb) in high-percentage alcohols [64], which were presented in the first part of the manuscript, decided also to check potential exceedances of heavy metals (Cd and Pb) in fake whisky samples. In the mentioned standards, the maximum lead content was set at 0.3 mg/L, and the cadmium one at 0.03 mg/L. This time, there were only exceedances in the case of cadmium. The exceedances of the maximum allowable concentrations concerned three samples (F10, F11 and F12), which came from a common source. The values recorded for Cd in these cases ranged from 32.25–65.90 μg/L.

3.2. Comparison of Elemental Profiles of Authentic and Counterfeit Whisky

In this experiment, a set of counterfeit and authentic samples was analyzed to reveal the possible differences between them, as well as to detect and identify the elemental fingerprint group of genuine and fake whisky. Apart from the above-mentioned 30 elements (Hg was omitted because its concentration in each sample was below the limit of quantification) and the pH value, in the analysis, the values of Sr and Cu isotope ratios were also used. These ratios were calculated based on the number of counts for each of the isotope as a result of the semi-quantitative analysis. In the case of Sr isotopes, the interference from Rb was corrected. For copper, an analysis was performed on the basis of the $^{63}Cu/^{65}Cu$ isotope ratio. In turn, for Sr, the following isotopic ratios were used: $^{88}Sr/^{86}Sr$, $^{84}Sr/^{86}Sr$, $^{87}Sr/^{86}Sr$, as these are the parameters most frequently used in food authentication [45].

On the basis of the Kruskal–Wallis test, the existence of statistically significant differences in the concentration of the following elements was demonstrated: Be, Ca, Cu, Li, Mg, Mo, S, Sn, Sr and pH value (Table 3). In all mentioned cases the level of significance (p) was less than 0.05.

Table 3. Contents of selected elements (with statistically significant differences) in the measured fake and authentic alcohol samples (n = 205) [μg/L].

Element	Code	N	Mean	Median	Min	Max	Std. Dev.
^{9}Be	A	170	0.100	0.092	<LOQ	0.300	0.050
	F	35	0.130	0.120	<LOQ	0.500	0.100
^{59}Co	A	170	4.530	2.468	0.406	74.90	7.870
	F	35	9.920	5.260	1.409	42.20	10.10
^{63}Cu	A	170	473.7	216.0	16.25	5252	736.4
	F	35	4021	56.86	1.922	33,212	7367
^{7}Li	A	170	21.36	12.27	0.474	399.5	35.40
	F	35	67.12	19.25	<LOQ	825.4	140.8
^{95}Mo	A	170	1.790	1.066	<LOQ	32.30	3.320
	F	35	11.07	1.590	<LOQ	108.4	30.30
^{60}Ni	A	170	24.01	12.96	3.201	301.3	33.68
	F	35	62.71	39.86	2.418	411.0	73.70
^{118}Sn	A	170	9.800	4.672	<LOQ	44.50	11.31
	F	35	13.89	9.810	<LOQ	34.70	11.00
^{88}Sr	A	170	47.18	45.81	15.84	119.2	19.80
	F	35	133.0	53.72	14.15	765.1	168.8
Ca 393.366	A	170	14.66	9185	723.8	175.4	17.98
	F	35	35.73	22.91	1994	271.1	50.19
Mg 279.553	A	170	1487	1046	208.5	11.55	13.93
	F	35	5370	1577	465.4	33.07	764
S 180.731	A	170	7126	4648	296.7	69.91	8654
	F	35	20.89	14.68	197.6	231.7	39.56
pH value	A	170	3.63	3.63	1.95	6.20	0.68
	F	35	4.71	4.39	2.79	8.70	1.50

Comparing the median values of the two groups under consideration (fake and authentic whisky samples) in each case, except for copper, higher values were noted for products with unidentified identity. Although the highest content of copper was recorded in the fake sample (33.21 μg/L), the median and mean values of the samples belonging to the group of authentic products were much higher. However, it should be noted that in the group of false samples there were five objects with a much higher concentration of copper. These were samples coded as F4 and F5 and from F9 to F11 with a copper content in the range from 12.89 to 33.21 μg/L. As emphasized in the first part of the work, the presence of copper in alcohol is undoubtedly related to the material of the apparatus used in the production process, and more specifically during distillation. Therefore, the alcohols coded as F4, F5, F9, F10 and F11 have most certainly been distilled in copper stills, resembling the high-quality single malt whisky. As it was underlined in the previous paper, differentiation of the authentic samples may be influenced by several overlapping parameters. Moreover, counterfeit samples can be of variable origin and composition, so it is impossible to characterize them as one group with identical or comparable attributes. When the influence of overlapping parameters was eliminated, in the case of authentic samples, the increasing concentration of V, Cr, Ni, Sr, Sb, Bi, Zn, Mg, K and P with the age of the analyzed samples was revealed (despite the lack of statistically significant differences). A similar result was recorded for the comparison of authentic and false objects in this study. Despite the lack of statistically significant differences, higher values of both the median and mean of Mn and P and the median value for K were recorded for the genuine samples, which were maturated (minimum 3 years). Thus, it is possible to clearly indicate the influence of aging on the levels of phosphorus and manganese and potassium, as these elements can be selected as markers for the identification of products with adulterated maturation. The chemical composition of wood is the explanation for the higher content of the above-mentioned elements in the authentic samples in relation to the false ones. Unadulterated whisky is matured in oak barrels, usually incinerated from the inside. The presence of phosphorus and potassium is directly related to the oxides formed during the firing of wooden barrels for aging alcohol. On the other hand, phosphorus, as a macroelement necessary for plant development, may accumulate in various parts of plants when migrating from the soil. The main form of phosphorus in soil is phosphates, including manganese phosphates [65,66]. In addition, manganese compounds are used as wood preservatives, which may also affect the content of this element in alcohol stored in oak barrels [67]. Thus, the longer the alcohol stays in contact with wooden barrels, the greater the migration of these elements into the product. It is true that the aforementioned average concentration of copper was higher in authentic samples, i.e., those subjected to the aging process, however, the content of this element should be associated with the equipment used for production rather than with the age parameter.

Among the elements listed in Table 3, for which the existence of statistically significant differences has been demonstrated, the presence of sulfur should be commented on. As reported in the literature data, sulfur volatile compounds generated during the whisky production process influence their quality to a large degree [68]. The selected alkyl sulfides (dimethyl sulfide (DMS), dimethyl disulfide (DMDS) and dimethyl trisulfide (DMTS)) have been recognized as age markers for whisky, as the level decreases with the time the alcohol spends in the barrel [69,70]. Comparing the mean and the median values of the groups of false and authentic samples, it is clear that the concentration of S in the set of counterfeit samples (not subjected to aging or with a falsified aging period) was an order of magnitude higher than in the original ones (which in each case were samples aged by at least 3 years). Thus, both the concentration of sulfur compounds, as evidenced by the literature, and the total sulfur content, as shown in this study, decrease with the aging of alcohol.

Also, the much higher pH value in the case of fake samples, as compared to the authentic ones, is worth emphasizing. This applies to both the mean and the median values. Although the set of authentic samples is much more numerous than the samples of unidentified identity, the pH values obtained in this group were much more similar and were in the acidic pH range. The counterfeit alcohol samples, on the other hand, had the pH ranging from 2.79 to 8.70, i.e., from acid to alkaline. Adherence to strict standards in the whisky production process ensures that certain physical and chemical parameters of alcohol are maintained within a given brand, including the characteristic pH value of the product. The large discrepancy in the results of the pH value in a small group of fake samples (including samples from a common source) suggests a lack of compliance with production standards and certain randomness during the production of this type of alcohol.

The comparison of the Cu and Sr isotope ratios of the genuine and false sample groups did not provide significant information allowing their better differentiation.

In the next step, the projection of cases on the factor plane for reduced data set was made. Since the significant influence of aging on the elemental profile of whisky had already been proven in earlier work, the age parameter was eliminated. Therefore, during the comparison of false and genuine samples, only the original samples were taken into account, which were aged for the legally required period (3 years).

As shown in Figure 1, quite a good separation between genuine and counterfeit samples using PCA was achieved. The vast majority of authentic samples are accumulated in one area of the graph (around the point of intersection of the coordinate axes), while the points belonging to the false samples are scattered over throughout the plot. This area contains over 70% of alcohol samples with unidentified identity. Despite the much smaller number of counterfeit samples, their large diversity in composition makes it impossible to characterize them as one group with similar physicochemical characteristics. Repeatedly, other authors have indicated that it is extremely difficult to find a marker occurring only in fake samples [42,71–75]. Most often, the problem arises from the type and nature of the adulterations. Depending on whether the adulteration concerns a lower alcohol content than the standard required [73] or on the addition of esters, aldehydes or organic acids [71,72], in order to reflect the age, taste, smell and quality of a given brand, a different and individual approach should be taken. Nonetheless, under such conditions, nontargeted screening followed by chemometric analysis can be a powerful instrument to uncover deviations from typical authentic whisky fingerprints.

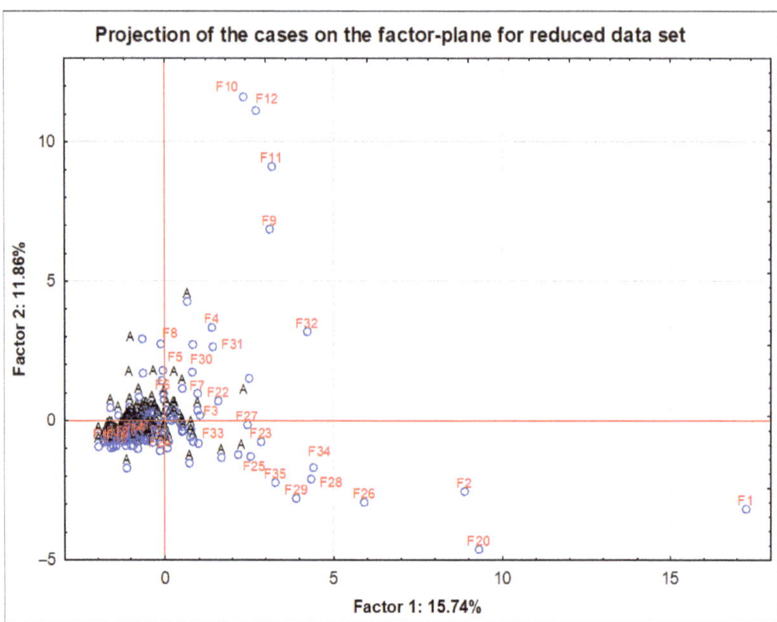

Figure 1. PCA score plot of 3-year-old authentic (A) and fake (F1–35) whisky samples.

Figure 1 resembles an analogous projection presented in the work of Stupak et al. [42]. The authors of the aforementioned work separated the samples of counterfeit and original whisky on the basis of selected markers measured with chromatographic techniques. In this case, in the PCA plot, all points belonging to the group of genuine products (both single malt and blended) were clustered in one common area, while objects belonging to the fake samples are dispersed across the graph.

3.3. *Counterfeit Whisky Analysis*

In the next steps, only samples marked as fake (35) were discussed separately with division to their sources of origin (1–9). On the basis of the Kruskal–Wallis test, the existence of statistically significant differences in the concentration of the following elements was demonstrated: B, Bi, Cd, Co, Fe, Mn, Mo, Ni, Pb, Sb and Zn. In each case, the level of significance (p) was less than 0.05. The most important statistical information connected with the division of fake samples against the sources is included in Table 4. It is worth noting that statistically significant differences for each of the elements, every time concerned, the source of the counterfeit whisky samples was marked as the number 2 (indicated as red color on Figure 2). Moreover, taking into account the median value for all elements listed in Table 5 (except Sn), the lowest concentrations were recorded for source 2.

Table 4. Groups with statistically significant differences.

Statistically Significant Differences	Elements
Source 6–Source 2	B
Source 3–Source 2	Fe; Mn; Mo; Sn
Source 1–Source 2	Bi; Cd; Co; Ni; Pb; Zn

Table 5. Contents of selected elements (with statistically significant differences) in the measured fake alcohol samples (n = 35) [μg/L].

Element	No. of Source	N	Mean	Median	Min	Max	Std. Dev.
^{11}B	1	7	2064	2238	<LOQ	3289	1055
	2	9	1803	1704	190.7	3758	1158
	3	6	2503	2728	1704	3059	657.0
	6	4	9078	8413	5970	13.52	3318
^{209}Bi	1	7	12.75	10.35	9.387	25.77	5.870
	2	9	<LOQ	<LOQ	<LOQ	<LOQ	<LOQ
	3	6	0.847	0.419	<LOQ	3.543	1.347
	6	4	2.300	2.705	<LOQ	3.790	1.618
^{111}Cd	1	7	26.05	11.06	3.204	65.90	28.11
	2	9	0.019	<LOQ	<LOQ	0.128	0.042
	3	6	2.690	2.166	<LOQ	7.325	2.857
	6	4	0.724	0.659	0.171	1.410	0.510
^{59}Co	1	7	26.09	23.90	13.77	42.21	9.000
	2	9	3.840	3.436	1.409	7.613	1.895
	3	6	7.218	5.584	3.698	12.53	4.056
	6	4	5.985	4.690	3.504	11.06	3.430
^{55}Mn	1	7	39.55	37.18	16.85	73.94	18.72
	2	9	14.12	5.192	2.377	64.16	20.26
	3	6	143.4	81.27	64.16	438.7	147.1
	6	4	49.89	31.59	6.299	130.0	58.44
^{95}Mo	1	7	1.280	1.560	<LOQ	2.130	0.840
	2	9	0.289	0.242	<LOQ	0.988	0.339
	3	6	56.68	59.58	1.982	108.4	56.76
	6	4	2.953	2.424	1.218	5.750	1.947
^{60}Ni	1	7	79.75	57.46	34.79	136.2	43.85
	2	9	13.77	10.21	2.419	30.11	10.04
	3	6	110.4	47.66	19.14	411.0	150.9
	6	4	75.36	74.30	69.34	83.49	6.114
^{208}Pb	1	7	29.65	30.75	22.99	35.60	4.430
	2	9	3.677	1.186	<LOQ	21.89	7.008
	3	6	10.72	13.38	3.569	13.42	4.308
	6	4	12.09	12.43	6.553	16.95	5.358
^{118}Sn	1	7	4.330	4.737	<LOQ	8.600	2.860
	2	9	17.09	19.97	9.310	20.41	4.627
	3	6	29.63	29.59	23.42	34.65	3.617
	6	4	8.257	4.651	<LOQ	23.73	10.55
Fe 238.204	1	7	49.00	47.75	<LOQ	90.67	33.67
	2	9	0.036	<LOQ	<LOQ	0.316	0.105
	3	6	669.7	233.2	<LOQ	2735	1035
	6	4	7.496	<LOQ	<LOQ	29.98	14.99
Zn 213.856	1	7	11.01	5668	4353	39.82	12.80
	2	9	90.47	0.144	<LOQ	429.3	152.5
	3	6	859.7	725.1	111.2	1891	704.8
	6	4	4189	77.29	<LOQ	16.60	8276

<LOQ—limit of quantification

In the analyzed set of fake samples, nine different, independent sources were distinguished and according to this criterion a division was made and what is worth mentioning is the fact that within these separated groups, alcohol samples of a completely different nature were observed. This means that they were produced by one manufacturer, but some of them are "raw" products, i.e., distillates that have not undergone any treatment to change their color or taste, whereas others are finished products intended for sale and consumption. However, the tendency that can be noticed in the projection of the cases on the factor plane for the fake products presented in Figure 2 is the grouping of samples within a common source. Each group has been marked with a different color. Sources 8 (F2) and 9 (F1) are represented by single samples. Sample F1 (source 9) is distinguished by the highest values of Li, Mn, Sr and Ba in relation to the other counterfeit samples, hence its extreme position on the graph presented below. Within the source 1 (marked in green), a cluster of items from F30 to F32 can be distinguished. These are samples of the same alcohol coming probably from one production batch but taken from three independent bottles. It should be mentioned that this alcohol has been enriched with wood extracts in order to give it the characteristic whisky aromas. The other samples in this group are of a completely different nature. Moreover, samples F9–12 and F30–32 contain the highest concentrations of Cd in the tested set of false ones. For items F10–12, the permissible level of this element has been exceeded. The samples from sources 2, 4, 5 and 7 in Figure 2 form the most central, individual clusters. An interesting group is consisted of the samples from source 3 marked in yellow in Figure 2. Points F3, F22 and F24 are samples of high-strength distillates. In turn, samples F20, F26 and F34 are flavored products, which are made from these distillates. They have been enriched with sugar and fruit juices. These products were supposed to resemble whisky-based fruit liqueurs.

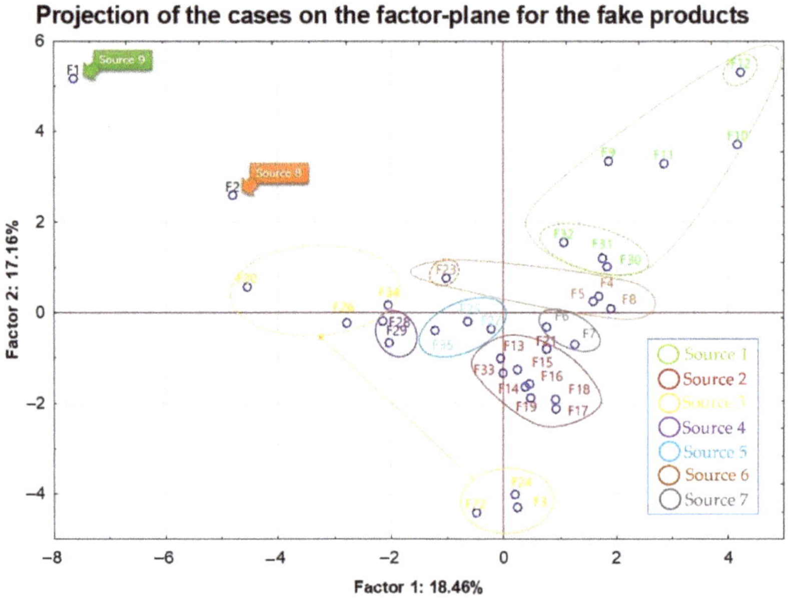

Figure 2. PCA score plot of fake whisky with division on 9 different sources of samples.

4. Conclusions

Mn, K and P are elements with higher concentrations recorded in the case of authentic samples. Their presence is directly related to the aging period of alcohol and can be indicated as markers for the identification of fraudulent activity in this respect. Another indicator certainly associated to the whisky maturation process in barrels is S. In products that were not aged or the aging period was much shorter than legally required (fake samples), the concentration of this element was much higher, compared to the original samples (in all cases subjected to the aging process). Counterfeit samples can be of variable origin and composition, so they cannot be characterized as one group with identical or comparable attributes. Often, the element of randomness dominates in the production of such alcohols. The use of unsuitable ingredients or production equipment, as well as inadequate knowledge in this field, cause the lack of repeatability of the taste and smell characteristics of alcohol beverages. This is evidenced by, for example, the failure to meet the standards for the maximum content of heavy metals in high-percentage alcohols. The adulteration of food products, including whisky, may be of various characters. It can refer to a reduced percentage of alcohol or the addition of various organic compounds to improve the visual and flavor properties. Therefore, the identification of the falsification of a different nature requires the use of a wide range of analytical techniques and often an individual approach.

The results presented in this article constitute the final part of a broad characteristic of the elemental composition carried out for 205 whisky samples. As our research revealed, the elemental analysis supported by statistical tools may provide beneficial information, especially in the context of the differentiation of alcohol samples in regard to such parameters as type, origin and detecting age-related adulteration of whisky.

Author Contributions: A.P., P.W. and M.G. performed an elemental analysis of all samples, analyzed the data, performed the chemometric analysis, prepared the paper; E.M. and J.A. methodological and technical support for research, performed an elemental analysis of all samples, prepared the paper; K.J.—conducted substantive supervision; M.I.S.-J. conducted substantive supervision and final review. All authors have read and agreed to the published version of the manuscript.

Funding: This research received no external funding.

Institutional Review Board Statement: Not applicable.

Informed Consent Statement: Not applicable.

Data Availability Statement: The date are available from the corresponding author.

Acknowledgments: The authors of this paper would like to thank the invaluable technical support Patrycja Skrzek.

Conflicts of Interest: The authors declare no conflict of interest.

References

1. Luykx, D.M.A.M.; van Ruth, S.M. An overview of analytical methods for determining the geographical origin of food products. *Food Chem.* **2008**, *107*, 897–911. [CrossRef]
2. Lakshmi, V. Food adulteration. *Int. J. Sci. Invent. Today* **2012**, *1*, 101–113.
3. Everstine, K. Economically Motivated Adulteration: Implications for Food Protection and Alternate Approaches to Detection. Ph.D. Thesis, University of Minnesota, Minneapolis, MN, USA, 2013.
4. Ellis, D.I.; Brewster, V.L.; Dunn, W.B.; Allwood, J.W.; Golovanov, A.P.; Goodacre, R. Fingerprinting food: Current technologies for the detection of food adulteration and contamination. *Chem. Soc. Rev.* **2012**, *41*, 5706. [CrossRef] [PubMed]
5. Tsagkaris, A.S.; Koulis, G.A.; Danezis, G.P.; Martakos, I.; Dasenaki, M.; Georgiou, C.A.; Thomaidis, N.S. Honey authenticity: Analytical techniques, state of the art and challenges. *RSC Adv.* **2021**, *11*, 11273. [CrossRef] [PubMed]
6. Makowicz, E.; Jasicka-Misiak, I.; Teper, D.; Kafarski, P. HPTLC Fingerprinting—Rapid Method for the Differentiation of Honeys of Different Botanical Origin Based on the Composition of the Lipophilic Fractions. *Molecules* **2018**, *23*, 1811. [CrossRef] [PubMed]
7. Gonzalvez, A.; Armenta, S.; de la Guardia, M. Trace-element composition and stable-isotope ratio for discrimination of foods with Protected Designation of Origin. *Trends Anal. Chem.* **2009**, *28*, 1295–1311. [CrossRef]
8. Collins, E.J.T. Food adulteration and food safety in Britain in the 19th and early 20th centuries. *Food Policy* **1993**, *18*, 95–109. [CrossRef]
9. Teletchea, F.; Maudet, C.; Hänni, C. Food and forensic molecular identification: Up-date and challenges. *Trends Biotechnol.* **2005**, *23*, 359–366. [CrossRef]
10. Mori, C.; Matsumura, S. Current issues for mammalian species identification in forensic science: A review. *Int. J. Legal Med.* **2021**, *135*, 3–12. [CrossRef]
11. Rossi Scalco, A.; Miller Devós Ganga, G.; Cristina De Oliveira, S.; Baker, G. Development and validation of a scale for identification of quality attributes of agri-food products in short chains. *Geoforum* **2020**, *111*, 165–175. [CrossRef]
12. Lim, J.H.; Bae, D.; Fong, A. Titanium dioxide in food products: Quantitative analysis using ICP-MS and Raman spectroscopy. *J. Agric. Food Chem.* **2018**, *66*, 13533–13540. [CrossRef]
13. Grassi, S.; Casiraghi, E.; Alamprese, C. Handheld NIR device: A non-targeted approach to assess authenticity of fish fillets and patties. *Food Chem.* **2018**, *243*, 382–388. [CrossRef]
14. Manfredi, M.; Robotti, E.; Quasso, F.; Mazzucco, E.; Calabrese, G.; Marengo, E. Fast classification of hazelnut cultivars through portable infrared spectroscopy and chemometrics. *Spectrochim. Acta A Mol. Biomol. Spectrosc.* **2018**, *189*, 427–435. [CrossRef]
15. Gajek, M.; Pawlaczyk, A.; Szynkowska-Jozwik, M.I. Multi-Elemental Analysis of Wine Samples in Relation to Their Type, Origin, and Grape Variety. *Molecules* **2021**, *26*, 214. [CrossRef]

16. Peng, C.; Zhang, Y.; Song, W.; Cai, H.; Wang, Y.; Granato, D. Characterization of Brazilian coffee based on isotope ratio mass spectrometry ($\delta^{13}C$, $\delta^{18}O$, $\delta^{2}H$, and $\delta^{15}N$) and supervised chemometrics. *Food Chem.* **2019**, *297*, 124963. [CrossRef]
17. Park, J.H.; Choi, S.H.; Bong, Y.S. Geographical origin authentication of onions using stable isotope ratio and compositions of C, H, O, N, and S. *Food Control* **2019**, *101*, 121–125. [CrossRef]
18. Barberis, E.; Amede, E.; Dondero, F.; Marengo, E.; Manfredi, M. New Non-Invasive Method for the Authentication of Apple Cultivars. *Foods* **2022**, *11*, 89. [CrossRef]
19. Datta, A.; Sharma, A.N.G.V. Quantification of minor and trace elements in raw and branded turmeric samples using instrumental neutron activation analysis utilizing apsara-U reactor for possible applications to forensic science. *J. Radioanal. Nucl. Chem.* **2020**, *325*, 967–975. [CrossRef]
20. Saadata, S.; Pandyaa, H.; Deya, A.; Rawtani, D. Food forensics: Techniques for authenticity determination of food products. *Forensic Sci. Int.* **2022**, *333*, 111243. [CrossRef]
21. Laursen, K.H.; Schjoerring, J.K.; Kelly, S.D.; Husted, S. Authentication of organically grown plants—Advantages and limitations of atomic spectroscopy for multi-element and stable isotope analysis. *TrAC Trends Anal. Chem.* **2014**, *59*, 73–82. [CrossRef]
22. Barbosa, R.M.; Nacano, L.R.; Freitas, R.; Batista, B.L.; Barbosa Jr, F. The use of decision trees and naïve bayes algorithms and trace element patterns for controlling the authenticity of free-range-pastured hens' eggs. *J. Food Sci.* **2014**, *79*, 1672–1677. [CrossRef]
23. Rodrigues, S.M.; Otero, M.; Alves, A.A.; Coimbra, J.; Coimbra, M.A.; Pereira, E.; Duarte, A.C. Elemental analysis for categorization of wines and authentication of their certified brand of origin. *J. Food Compos. Anal.* **2011**, *24*, 548–562. [CrossRef]
24. Farmaki, E.G.; Thomaidis, N.S.; Minioti, K.S.; Ioannou, E.; Georgiu, C.A.; Efstathiou, C.E. Geographical characterization of Greek olive oils using rare earth elements content and supervised chemometric techniques. *Anal. Lett.* **2012**, *45*, 920–932. [CrossRef]
25. Baroni, M.V.; Podio, N.S.; Badini, R.; Inga, C.M.; Ostera, H.; Cagnoni, M.; Gautier, E.A.; Peral-Garcia, P.; Hoogewerff, J.; Wunderlin, D.A. Linking soil, water, and honey composition to assess the geo-graphical origin of Argentinean honey by multielemental and isotopic analyses. *J. Agric. Food Chem.* **2015**, *63*, 4638–4645. [CrossRef]
26. Barbosa, R.M.; Batista, B.L.; Varrique, R.M.; Coelho, V.A.; Campiglia, A.D.; Barbosa, F., Jr. The use of advanced chemometric techniques and trace element levels for controlling the authenticity of organic coffee. *Food Res. Int.* **2014**, *61*, 246–251. [CrossRef]
27. Ma, G.; Zhang, Y.; Zhang, J.; Wang, G.; Chen, L.; Zhang, M.; Liu, T.; Liu, X.; Lu, C. Determining the geographical origin of Chinese green tea by linear discriminant analysis of trace metals and rare earth elements: Taking *Dongting Biluochun* as an example. *Food Control* **2016**, *59*, 714–720. [CrossRef]
28. Camin, F.; Wehrens, R.; Bertoldi, D.; Bontempo, L.; Ziller, L.; Perini, M.; Nicolini, G.; Nocetti, M.; Larcher, R. H, C, N and S stable isotopes and mineral profiles to objectively guarantee the authenticity of grated hard cheeses. *Anal. Chim. Acta* **2012**, *711*, 54–59. [CrossRef]
29. Drivelos, S.A.; Higgins, K.; Kalivas, J.H.; Haroutounian, S.A.; Georgiou, C.A. Data fusion for food authentication. Combining rare earth elements and trace metals to discriminate "Fava Santorinis" from other yellow split peas using chemometric tools. *Food Chem.* **2014**, *165*, 316–322. [CrossRef]
30. Danezis, G.P.; Tsagkaris, A.S.; Brusic, V.; Georgiou, C.A. Food authentication: State of the art and prospects. *Curr. Opin. Food Sci.* **2016**, *10*, 22–31. [CrossRef]
31. Ibáñez, C.; García-Cañas, V.; Valdés, A.; Simó, C. Novel MS-based approaches and applications in food metabolomics. *Trends Anal. Chem.* **2013**, *52*, 100–111. [CrossRef]
32. Bat, K.B.; Vodopivec, B.M.; Eler, K.; Ogrinc, N.; Mulič, I.; Masuero, D.; Vrhovšek, U. Primary and secondary metabolites as a tool for differentiation of apple juice according to cultivar and geographical origin. *LWT* **2018**, *90*, 238–245. [CrossRef]
33. Guo, J.; Yuan, Y.; Dou, P.; Yue, T. Multivariate statistical analysis of the polyphenolic constituents in kiwifruit juices to trace fruit varieties and geographical origins. *Food Chem.* **2017**, *232*, 552–559. [CrossRef] [PubMed]
34. Sagratini, G.; Maggi, F.; Caprioli, G.; Cristalli, G.; Ricciutelli, M.; Torregiani, E.; Vittori, S. Comparative study of aroma profile and phenolic content of montepulcano monovarietal red wines from the marches and abruzzo regions of Italy using HS-SPME–GC–MS and HPLC–MS. *Food Chem.* **2012**, *132*, 1592–1599. [CrossRef] [PubMed]
35. Liu, J.; Chen, N.; Yang, J.; Yang, B.; Ouyang, Z.; Wu, C.; Yuan, Y.; Wang, W.; Chen, M. An integrated approach combining HPLC, GC/MS, NIRS, and chemometrics for the geographical discrimination and commercial categorization of saffron. *Food Chem.* **2018**, *253*, 284–292. [CrossRef]
36. Opatić, A.M.; Nečemer, M.; Lojen, S.; Masten, J.; Zlatić, E.; Šircelj, H.; Stopar, D.; Vidrih, R. Determination of geographical origin of commercial tomato through analysis of stable isotopes, elemental composition and chemical markers. *Food Control* **2018**, *89*, 133–141. [CrossRef]
37. Yudthavorasit, S.; Wongravee, K.; Leepipatpiboon, N. Characteristic fingerprint based on gingerol derivative analysis for discrimination of ginger (Zingiber officnale) according to geographical origin using HPLC-DAD combined with chemometrics. *Food Chem.* **2014**, *158*, 101–111. [CrossRef]
38. Wisniewska, P.; Sliwinska, M.; Namiesnik, J.; Wardencki, W.; Dymerski, T. The Verification of the Usefulness of Electronic Nose Based on Ultra-Fast Gas Chromatography and Four Different Chemometric Methods for Rapid Analysis of Spirit Beverages. *J. Anal. Methods Chem.* **2016**, *5*, 1–12. [CrossRef]
39. Heinz, H.A.; Elkins, J.T. Comparison of unaged and barrel aged whiskies from the same Mash Bill using gas chromatography/mass spectrometry. *J. Brew. Distill.* **2019**, *8*, 1–6.

40. Daute, M.; Jack, F.; Baxter, I.; Harrison, B.; Grigor, J.; Walker, G. Comparison of Three Approaches to Assess the Flavour Characteristics of Scotch Whisky Spirit. *Appl. Sci.* **2021**, *11*, 1410. [CrossRef]
41. Wisniewska, P.; Boqué, R.; Borràs, E.; Busto, O.; Wardencki, W.; Namieśnik, J.; Dymerski, T. Authentication of whisky due to its botanical origin and way of production by instrumental analysis and multivariate classification methods. *Spectrochim. Acta Part A Mol. Biomol. Spectrosc.* **2017**, *173*, 849–853. [CrossRef]
42. Stupak, M.; Goodall, I.; Tomaniova, M.; Pulkrabova, J.; Hajslova, J. A novel approach to assess the quality and authenticity of Scotch Whisky based on gas chromatography coupled to high resolution mass spectrometry. *Anal. Chim. Acta* **2018**, *1042*, 60–70. [CrossRef]
43. Cvetkovic, D.; Stojilkovic, P.; Zvezdanovic, J.; Stanojevic, J.; Stanojevic, L.; Karabegovic, I. The identification of volatile aroma compounds from local fruit based spirits using a headspace solid-phase microextraction technique coupled with the gas chromatography-mass spectrometry. *Adv. Technol.* **2020**, *9*, 19–28. [CrossRef]
44. Fortunato, G.; Mumic, K.; Wunderli, S.; Pillonel, L.; Bosset, J.O.; Gremaud, G. Application of strontium isotope abundance ratios measured by MC-ICP-MS for food authentication. *J. Anal. At. Spectrom.* **2004**, *19*, 227–234. [CrossRef]
45. Kelly, S.; Heaton, K.; Hoogewerff, J. Tracing the geographical origin of food: The application of multi-element and multi-isotope analysis. *Trends Food Sci. Tech.* **2005**, *16*, 555–567. [CrossRef]
46. Manca, G.; Franco, M.A.; Versini, G.; Camin, F.; Rossmann, A.; Tola, A. Correlation between multielement stable isotope ratio and geographical origin in Peretta cows' milk cheese. *J. Dairy Sci.* **2006**, *89*, 831. [CrossRef]
47. Longobardi, F.; Casiello, G.; Ventrella, A.; Mazzilli, V.; Nardelli, A.; Sacco, D.; Catucci, L.; Agostiano, A. Electronic nose and isotope ratio mass spectrometry in combination with chemometrics for the characterization of the geographical origin of Italian sweet cherries. *Food Chem.* **2015**, *170*, 90–96. [CrossRef]
48. Longobardi, F.; Casiello, G.; Cortese, M.; Perini, M.; Camin, F.; Catucci, L.; Agostiano, A. Discrimination of geographical origin of lentils (*Lens culinaris* Medik.) using isotope ratio mass spectrometry combined with chemometrics. *Food Chem.* **2015**, *188*, 343–349. [CrossRef]
49. De Rijke, E.; Schoorl, J.C.; Cerli, C.; Vonhof, H.B.; Verdegaal, S.J.A.; Vivó-Truyols, G.; Lopatka, M.; Dekter, R.; Bakker, D.; Sjerps, M.J.; et al. The use of δ^2H and δ^{18}O isotopic analyses combined with chemometrics as a traceability tool for the geographical origin of bell peppers. *Food Chem.* **2016**, *204*, 122–128. [CrossRef]
50. Luo, D.; Dong, H.; Luo, H.; Xian, Y.; Wan, J.; Guo, X.; Wu, Y. The application of stable isotope ratio analysis to determine the geographical origin of wheat. *Food Chem.* **2015**, *174*, 197–201. [CrossRef]
51. Fan, S.; Zhong, Q.; Gao, H.; Wang, D.; Li, G.; Huang, Z. Elemental profile and oxygen isotope ratio (δ^{18} O) for verifying the geographical origin of Chinese wines. *J. Food Drug Anal.* **2018**, *26*, 1033–1044. [CrossRef]
52. Dinca, O.R.; Ionete, R.E.; Costinel, D.; Geana, I.E.; Popescu, R.; Stefanescu, I.; Radu, G.L. Regional and vintage discrimination of romanian wines based on elemental and isotopic fingerprinting. *Food Anal. Methods.* **2016**, *9*, 2406–2417. [CrossRef]
53. Ciepielowski, G.; Pacholczyk-Sienicka, B.; Fraczek, T.; Klajman, K.; Paneth, P.; Albrecht, L. Comparison of quantitative NMR and IRMS spectrometry for the authentication of "Polish Vodka". *J. Sci. Food Agric.* **2019**, *99*, 263–268. [CrossRef] [PubMed]
54. Aylott, R. Chapter 16—Analytical strategies supporting protected designations of origin for alcoholic beverages. In *Comprehensive Analytical Chemistry*, 2nd ed.; de la Guardia, M., Gonzálvez, A., Eds.; Elsevier: London, UK, 2013; Volume 60, pp. 409–438. [CrossRef]
55. Reid, K.J.G.; Swan, J.S.; Gutteridge, C.S. Assessment of Scotch whisky quality by pyrolysis-mass spectrometry and the subsequent correlation of quality with the oak wood cask. *J. Anal. Appl. Pyrolysis* **1993**, *25*, 49–62. [CrossRef]
56. Wisniewska, P.; Dymerski, T.; Wardencki, W.; Namiesnik, J. Chemical composition analysis and authentication of whisky. *J. Sci. Food Agric.* **2015**, *95*, 2159–2166. [CrossRef]
57. Campo, E.; Cacho, J.; Ferreira, V. Solid phase extraction, multidimensional gas chromatography mass spectrometry determination of four novel aroma powerful ethyl esters: Assessment of their occurrence and importance in wine and other alcoholic beverages. *J. Chromatogr. A* **2007**, *1140*, 180–188. [CrossRef]
58. Campillo, N.; Peñalver, R.; Hernández-Córdoba, M. Solid-phase microextraction for the determination of haloanisoles in wines and other alcoholic beverages using gas chromatography and atomic emission detection. *J. Chromatogr. A* **2008**, *1210*, 222–228. [CrossRef]
59. Barbeira, P.J.S.; Stradiotto, N.R. Anodic stripping voltammetric determination of Zn, Pb and Cu traces in whisky samples. *Fresen. J. Anal. Chem.* **1998**, *361*, 507–509. [CrossRef]
60. Adam, T.; Duthie, E.; Feldmann, J. Investigations into the use of copper and other metals as indicators for the authenticity of Scotch whiskies. *J. Inst. Brew.* **2002**, *108*, 459–464. [CrossRef]
61. Gajek, M.; Pawlaczyk, A.; Jozwik, K.; Szynkowska-Jozwik, M.I. The Elemental Fingerprints of Different Types of Whisky as Determined by ICP-OES and ICP-MS Techniques in Relation to Their Type, Age, and Origin. *Foods* **2022**, *11*, 1616. [CrossRef]
62. Shand, C.A.; Wendler, R.; Dawson, D.; Yates, K.; Stephenson, H. Multivariate analysis of Scotch whisky by total reflection x-ray fluorescence and chemometric methods: A potential tool in the identification of counterfeits. *Anal. Chim. Acta* **2017**, *976*, 14–24. [CrossRef]
63. Pawlaczyk, A.; Gajek, M.; Jozwik, K.; Szynkowska, M.I. Multielemental Analysis of Various Kinds of Whisky. *Molecules* **2019**, *24*, 1193. [CrossRef]

64. Regulation of the Minister of Health on the Maximum Levels of Biological and Chemical Contaminants that May Be Present in Food, Food Ingredients, Permitted Additives, Processing Aids or on the Surface of Food of 13 January 2003 (Journal of Laws of 2003). Available online: Sejm.gov.pl (accessed on 10 December 2021).
65. Phosphorus, Soil Phosphorus. *Forest Encyclopedia*. Available online: https://www.encyklopedialesna.pl/haslo/fosfor-fosfor-glebowy/ (accessed on 12 May 2022).
66. Jama, A.; Nowak, W. Uptaking of Macronutrients from the Soil Covered with Sewage Sludge by Willow (*Salix viminalis* L.) and Its Hybrids. *Nauka Przyr. Technol.* **2011**, *5*, 6.
67. Surgiewicz, J. Mangan i jego związki. *Podstawy I Metod. Oceny Sr. Pr.* **2012**, *1*, 111–116.
68. Wanikawa, A.; Sugimoto, T. A Narrative Review of Sulfur Compounds in Whisk(e)y. *Molecules* **2022**, *27*, 1672. [CrossRef]
69. Leppänen, O.; Ronkainen, P.; Denslow, J.; Laakso, R.; Lindeman, A.; Nykanen, I. Polysulphides and thiophenes in whisky. In *Flavour of Distilled Beverages: Origin and Development*; Piggott, J.R., Ed.; E. Horwood Ltd.: Chichester, UK, 1983; pp. 206–214.
70. Leppänen, O.; Denslow, J.; Ronkainen, P. A gas chromatographic method for the accurate determination of low concentrations of volatile sulphur compounds in alcoholic beverages. *J. Inst. Brew.* **1979**, *85*, 350–353. [CrossRef]
71. Ashok, P.C.; Praveen, B.B.; Dholakia, K. Near infrared spectroscopic analysis of single malt Scotch whisky on an optofluidic chip. *Opt. Express* **2011**, *19*, 22982–22992. [CrossRef]
72. Backhaus, A.; Ashok, P.C.; Praveen, B.B.; Dholakia, K.; Seiffert, U. Classifying Scotch whisky from near-infrared Raman spectra with a radial basis function network with relevance learning. In Proceedings of the 20th European Symposium on Artificial Neural Networks, Bruges, Belgium, 25–27 April 2012.
73. McIntyre, A.C.; Bilyk, M.L.; Nordon, A.; Colquhoun, G.; Littlejohn, D. Detection of counterfeit Scotch whisky samples using mid-infrared spectrometry with an attenuated total reflectance probe incorporating polycrystalline silver halide fibres. *Anal. Chim. Acta* **2011**, *690*, 228–233. [CrossRef]
74. Meier-Augenstein, W.; Kemp, H.F.; Hardie, S.M.L. Detection of counterfeit Scotch whisky by ^{2}H and ^{18}O stable isotope analysis. *Food Chem.* **2012**, *133*, 1070–1074. [CrossRef]
75. Garcia, J.S.; Vaz, B.G.; Corilo, Y.E.; Ramires, C.F.; Saraiva, S.A.; Sanvido, G.B.; Schmidt, E.M.; Maia, D.R.J.; Cosso, R.G.; Zacca, J.J.; et al. Whisky analysis by electrospray ionization-Fourier transform mass spectrometry. *Food Res. Int.* **2013**, *51*, 98–106. [CrossRef]

Article

Pattern Recognition Approach for the Screening of Potential Adulteration of Traditional and Bourbon Barrel-Aged Maple Syrups by Spectral Fingerprinting and Classical Methods

Kuanrong Zhu [1], Didem P. Aykas [2] and Luis E. Rodriguez-Saona [1,*]

[1] Department of Food Science and Technology, The Ohio State University, 110 Parker Food Science and Technology 2015 Fyffe Road, Columbus, OH 43210, USA; zhu.1421@buckeyemail.osu.edu
[2] Department of Food Engineering, Faculty of Engineering, Adnan Menderes University, Aydin 09100, Turkey; didem.cinkilic@adu.edu.tr
* Correspondence: rodriguez-saona.1@osu.edu; Tel.: +1-614-2923339

Abstract: This study aims to generate predictive models based on mid-infrared and Raman spectral fingerprints to characterize unique compositional traits of traditional and bourbon barrel (BBL)-aged maple syrups, allowing for fast product authentication and detection of potential ingredient tampering. Traditional (n = 23) and BBL-aged (n = 17) maple syrup samples were provided by a local maple syrup farm, purchased from local grocery stores in Columbus, Ohio, and an online vendor. A portable FT-IR spectrometer with a triple-reflection diamond ATR and a compact benchtop Raman system (1064 nm laser) were used for spectra collection. Samples were characterized by chromatography (HPLC and GC-MS), refractometry, and Folin–Ciocalteu methods. We found the incidence of adulteration in 15% (6 out of 40) of samples that exhibited unusual sugar and/or volatile profiles. The unique spectral patterns combined with soft independent modeling of class analogy (SIMCA) identified all adulterated samples, providing a non-destructive and fast authentication of BBL and regular maple syrups and discriminated potential maple syrup adulterants. Both systems, combined with partial least squares regression (PLSR), showed good predictions for the total °Brix and sucrose contents of all samples.

Keywords: maple syrups; adulteration; FT-IR; Raman; GC-MS; bourbon barrel aged

Citation: Zhu, K.; Aykas, D.P.; Rodriguez-Saona, L.E. Pattern Recognition Approach for the Screening of Potential Adulteration of Traditional and Bourbon Barrel-Aged Maple Syrups by Spectral Fingerprinting and Classical Methods. *Foods* **2022**, *11*, 2211. https://doi.org/10.3390/foods11152211

Academic Editor: Daniel Cozzolino

Received: 30 June 2022
Accepted: 19 July 2022
Published: 25 July 2022

1. Introduction

Native Americans are widely recognized as the first to discover the sweet sap dripping from the broken bark of sugar maple (Acer saccharum), which is the only ingredient of natural maple syrup products [1,2]. Maple syrup has a reputation for being a nutritious, classical sweetener and having a unique taste and flavor. According to the United States Department of Agriculture (USDA), the US production of maple syrup in 2019 totaled 4.37 million gallons with an estimated value of USD 135 million [3].

Pure maple syrups produced in North America comprise 68 ± 4% sucrose, 0.43 ± 1.11% glucose, 0.30 ± 0.54% fructose, a small amount of amino acids, various phenolic compounds, a trace amount of organic acids, including malic and fumaric acids, minerals, and salts [4]. Maple syrup is superior to other sweeteners because of its rich phenolic and phytohormone contents, which possess antioxidant properties and produce low glycemic and insulinemic responses [5].

Barrel aging is a process in which wine or spirits are stored and aged in wooden barrels. Chemical reactions take place during the aging process in which wine or spirits are absorbed into the wood constituents, including volatile compounds that contribute to the smell property and non-volatile compounds that correlate with color and mouthfeel properties [6]. In recent years, aging maple syrup in bourbon barrels has become popular and creates more value than traditional pure maple syrup. Bourbon barrel (BBL)-aged maple syrup is produced by aging traditional maple syrup in oak bourbon barrels for several weeks to months to develop richer bourbon flavor without adding any other ingredients [7]. In addition, in the process of aging, it is crucial to control the strength of extracted bourbon flavor to be neither too weak nor too strong to overshadow the maple flavor [7].

Maple syrup manufacturing is rather costly since it is regulated by law, specifying that the only ingredient in maple syrup is the maple sap [1]. Although sap contents may vary from different maple trees, in general, 1 L of traditional maple syrup is produced by concentrating around 35 L of maple sap to 66 °Brix [1]. Therefore, maple syrup could be potentially adulterated by adding inexpensive cane, beets, or corn syrup to the boiling sap or by blending the maple syrup with corn syrup due to financial incentives [1]. Since the taste of a small amount of cane sugar or corn syrup added to maple syrup is almost undetectable, the inclination to increase yields by fraudulent means can be substantial [8]. In addition, the even higher price of BBL-aged maple syrups may prompt the potential counterfeit of the aging process by having a minimum aging activity or by using an unqualified aging barrel that does not contain adequate bourbon residuals.

Traditional authentication methods, including chromatography, mass spectrometry, and stable isotope ratio analysis, have been applied to maple syrup studies [4,9]. Stuckel and Low developed a methodology to fingerprint oligosaccharides in maple syrup and to detect adulteration of high-fructose corn syrup and beet medium invert sugar via anion-exchange HPLC [10]. Carro et al. [9] authenticated maple syrup samples by using carbon stable isotope ratio analysis. An improved method of stable carbon isotope ratio mass spectroscopy was established by Tremblay and Paquin [11] with the isolation of malic acid to detect the addition of beet and cane sugar in maple syrup.

However, these methods are time-consuming and cost-prohibitive for most maple syrup manufacturers due to the requirements of expensive instrumentation and trained personnel [12]. Advances in the miniaturization of vibrational spectroscopy instruments combined with powerful chemometrics can overcome those problems by offering fast product authentication, non-destructive and real-time analysis [12]. Fourier transform infrared spectroscopy (FT-IR) is a vibrational spectroscopy technique that measures the absorbance and transmittance of infrared light. Raman spectroscopy is another type of vibrational spectroscopy using an intense light beam, such as a laser, to excite the sample molecules by inducing Raman-active vibrational modes and measuring inelastically scattered photons [13]. Since FT-IR measures the absorption of light, it is effective in measuring colored and fluorescence samples. At the same time, the presence of fluorescence creates optical noise for Raman measurements, which easily obscures the spectral fingerprint of the sample [14]. In addition, Raman scattering is based on polarizability changes in functional groups during molecule vibration [15]. Therefore, nonpolar bonds tend to give an intense Raman signal, while water in samples could be virtually disregarded due to a weak Raman signal [15]. Based on the reasons above, FT-IR and Raman are often used as complementary technologies for broader chemical identification. However, limited studies have employed vibrational spectroscopy for the authentication of maple syrups, and there is a gap in knowledge on the performance of portable/handheld devices for the detection of adulteration in maple syrups. Paradkar and others [8] reported the use of benchtop FT-IR, NIR, and FT-Raman systems to detect corn syrup adulteration in maple syrup. Mellado-Mojica and others [16] used FT-IR to contrast the carbohydrate composition of maple syrups against other sweeteners. In addition, chemometrics or multivariate analysis techniques have been proven to be successfully applied in the study of food matrices [17,18].

The objective of this research was to evaluate portable mid-infrared and Raman devices in generating predictive models for the non-destructive and fast fingerprinting of traditional and BBL maple syrups, allowing for product authentication and detection of potential ingredient tampering. This is the first study that characterizes a premium maple syrup aged in oak bourbon barrels, as there is no standard of identity or any other study reporting on this novel product. The use of miniaturized vibrational spectroscopies in maple syrup authentication can provide the industry with field-deployable devices for quality control and for preventing adulteration with cheaper ingredients.

2. Materials and Methods

2.1. Samples

Traditional and BBL maple syrup samples were kindly provided by a local maple syrup farm in Jefferson, OH (n = 12 (traditional), n = 8 (BBL)) and were purchased from local grocery stores in Columbus, OH (n = 7 (traditional), n = 5 (BBL)) that consisted of traditional maple syrups (n = 19), including dark, amber, and golden grades, and BBL-aged maple syrup (n = 13). In addition, table syrups (corn n = 2, cane n = 2, and mixture, consisting of cane, maple, and agave syrups n = 1) (n = 5) were purchased from grocery stores in Columbus, OH, USA for generating training models. An independent external validation set, consisting of traditional (n = 4) and BBL (n = 4) maple syrups, was purchased from an online vendor (Amazon.com, Inc. Seattle, WA, USA). All samples were stored in the refrigerator at 4°C and were equilibrated at room temperature before spectroscopic measurements and reference analyses.

2.2. Reference Analyses

2.2.1. °Brix

°Brix of each sample was measured with the heat-controlled refractometer (RX 5000i ATAGO, Bellevue, WA, USA). The syrup sample (~0.3 mL) was carefully pipetted onto the prism of the refractometer without creating any air bubbles, and measurement at 22 °C was recorded.

2.2.2. High-Performance Liquid Chromatography

Concentrations of sucrose, fructose, and glucose were measured with high-performance liquid chromatography (HPLC) (Shimadzu, Columbia, MD, USA). The HPLC was equipped with a SIL-20AHT autosampler, a CTO-20A oven, an LC-6AD pump, a CBM-20A controller, and a RID-10A refractive index detector. The syrup sample (~0.5 g) was weighed into a 15 mL centrifuge tube and diluted with (~7 mL) HPLC grade water. The actual weights of syrup and water were recorded. The mixture was vortexed for 40 sec and was filtered through the 0.2 μm filter (Phenomenex®, Torrance, CA, USA), and then filled into an HPLC vial. Isolated sugars were segregated by a Rezex RCM-Monosaccharide Ca+ 300 × 7.8 mm column (Phenomenex®). Sugars were eluted under the isocratic condition at 80 °C, using HPLC grade water as a mobile phase at a 1 mL/min flow rate for 20 min. LC Solutions software (Version 3.0, Shimadzu, Columbia, MD, USA) was used to integrate chromatograms automatically. The standard curve with concentration ranges from 10 to 50 mg/mL (>99% purity, Fisher Scientific, Fair Lawn, NJ, USA) was plotted to calculate each sugar content.

2.2.3. Total Phenolics

Total phenolic contents of maple syrups were determined with Folin–Ciocalteu (FC) method described by Waterhouse with some modification [19]. The syrup sample (~0.8 g) was weighed into a microcentrifuge tube and diluted with deionized (DI) water (~0.4 mL). The actual weights of syrup and water were recorded, and the diluted sample was vortexed for 40 s. The diluted sample (50 μL) was pipetted into a 96-well plate, followed by 200 μL DI water and 20 μL FC reagent. The mixture was mixed thoroughly by pipetting and incubated for 7 min at room temperature. The sodium carbonate solution (100 μL) was added to the mixture and incubated for 2 h under dark conditions at room temperature. The equilibrated sample's absorbance was measured at 765 nm. A standard curve constructed with gallic acid standard with concentration ranges from 125 to 800 μg/mL was used to quantify total phenolics. Results were expressed as micrograms of gallic acid equivalent (GAE) per 1 mL of distilled water.

2.2.4. Gas Chromatography—Mass Spectrometry

The volatile composition of the samples was identified using gas chromatography–mass spectrometry (GC-MS) (Agilent 7820A GC connected to a 5977B MS, Agilent Technologies, Santa Clara, CA, USA). A total of 1 g maple syrup sample was placed into a 20 mL clear screw-tread glass headspace vial (Restek, Bellefonte, PA, USA), and the vial was sealed with an 18 mm screw-tread PTFE/silicone septa vial cap (Restek, Bellefonte, PA, USA). The vial with the sample was placed onto a heating plate at 40 °C for 30 min to equilibrate the volatile compounds. A preconditioned SPME fiber (50/30 μm DVB/CAR/PDMS coated) (Supelco, Sigma-Aldrich, Bellefonte, PA, USA) assembly was inserted in the vial through the septa of the cap, and the volatiles were trapped by the fiber for 15 min. After the trapping, fiber assembly was removed from the vial and directly inserted through the GC-MS injection port. Compounds were desorbed at 250 °C for 1 min in splitless mode, followed by a 30 s purge flow (50 mL/min) to clean the fiber. A quality control (QC) sample was prepared by pooling 100 μL of each sample to monitor the performance of the method and identify qualified peaks. 2,3-hexanedione (Sigma-Aldrich, St. Louis, MO, USA) was prepared at 10 ppm concentration with distilled water and used as an internal standard (IS) to correct the variation through the run. A 40 μL of IS solution was added to each sample. The volatile compounds were separated on a DB-Wax column (60 m × 250 μm × 0.25 μm) (Agilent Technologies, Santa Clara, CA, USA). The oven was held at 60 °C for 5 min then ramped to 130 °C at 5 °C/min. This was followed by the second ramp of 5 °C/min to 240 °C, where it was held for 8 min. The MS acquisition was performed in scan mode between masses 25–300 m/z at a 2.7 scans/s rate. Data were extracted in the Agilent Masshunter Quantitative Analysis software. The spectral background was corrected, and only peaks that had a signal-to-noise (S/N) ratio higher than the detection limit (S/N > 5) were conserved. All compounds were tentatively identified using the NIST 14.L database by a Mass Spectral Library search.

2.2.5. Statistics of Reference Analysis

All reference laboratory analyses were performed in duplicate, and their range, minimum, maximum, mean, and standard deviation (SD) are determined. In addition, the standard error of laboratory (SEL) was calculated according to the method of Kovalenko et al. [20].

2.3. Vibrational Spectroscopy

2.3.1. Mid-Infrared Analysis

The mid-infrared data were collected with portable FT-IR spectroscopy (Agilent Technologies, Santa Clara, CA, USA) attached with a triple-reflection diamond Attenuated Total Reflectance (ATR) crystal. The ATR crystal has a sampling surface of 2 mm diameter and a 200 μm active area and provides ~6 μm depth of penetration. In addition, the FT-IR system is also attached with a deuterated triglycine sulfate (DTGS) detector and a Zinc Selenide (ZnSe) beam splitter. Spectra were collected from 4000 to 700 cm^{-1} with a resolution of 4 cm^{-1}. Sixty-four spectra were co-added in each sample collection to increase the signal-to-noise ratio. A spectral background was taken in between every measurement to reduce the environmental changes. Approximately 0.2 g of syrup sample was directly applied to the sampling surface of the ATR crystal, confirming that full coverage of the sample was achieved. Spectra for each sample were collected in triplicate, and collected spectral data were documented by Agilent MicroLab PC software (Agilent Technologies, Santa Clara, CA, USA).

2.3.2. Raman Analysis

About 3 mL of syrup sample was filled in a quartz cuvette (Hellma Analytics, Mulheim, Germany) with a 10 mm light path and measured with a compact benchtop Raman spectrometer WP 1064 (Wasatch Photonics, Durham, NC, USA). The Raman spectrometer was coupled with a laser operating at 1064 nm and an Indium Gallium Arsenide (InGaAs) detector. Spectra were collected from 350 to 1500 cm^{-1} with a resolution of 4 cm^{-1}. In addition, three scans were co-added and averaged to increase the signal-to-noise ratio of the spectrum, which has an integration time of 3 s. A spectral background was taken in between every measurement to reduce the environmental changes. The spectral collection was performed in triplication for each sample, and collected spectral data were documented by EnlightenTM software (Wasatch Photonics, Durham, NC, USA).

2.4. Multivariate Data Analysis

FT-IR and Raman spectra were exported and analyzed using Pirouette® multi-variate statistical analysis software (version 4.5, Infometrix Inc., Bothell, WA, USA). The mean spectrum of the three replicates was used for the statistical analysis. The collected FT-IR and Raman data were preprocessed with mean-centering to reduce micro multicollinearity and transformed with the Savitzky–Golay (SG) algorithm (35-point polynomial filter) in soft independent modeling of class analogy (SIMCA) and partial least squares regression (PLSR) models [21]. The SG algorithm was used to resolve overlapping spectroscopic signals and to improve their properties, also surpassing the instrument noise [22]. A 35-point smoothing filter was found as an optimal window length for our data set. The optimum window length was chosen to resolve essential details in the collected spectra and lessen signal noise. Mean centering and SG algorithms were chosen after evaluating the preprocessing quality of the spectral data with other options, including smoothing, normalization, and divide by; however, they were all outperformed by the combination of mean-centering and SG. An additional data transformation step of normalization (2-norm × 100) was applied in the case of PLSR analysis.

2.4.1. SIMCA

Classification analyses of maple syrups were performed using a supervised pattern recognition classification method SIMCA, which uses the previous understanding of the category membership of samples to classify new unrevealed samples in one of the known classes based on the pattern of measurements [23]. The cross-validation (leave-out-out) was used to assess the performance of the training model by analyzing the misclassification and generalization error [24]. The performance of the SIMCA model was also assessed with class projections, discriminating power, misclassification, and interclass distances (ICD), which interpret the quantitative similarity or dissimilarity of different classes and are widely accepted as samples that can be well differentiated when ICD >3 [24].

2.4.2. PLSR

The quantitative PLSR method was used for developing predictive training models of °Brix and sucrose contents by combining features from multiple linear regression and PCA. Cross-validation (leave-one-out) was used for internal validation of the training model. All syrup samples (n = 37) were randomly separated into calibration (~80% of the total samples) and external validation (~20% of the total samples) sets to evaluate the robustness of the trained models. Triplications of the same sample were used either in the training set or in the external validation set. The performance of the PLSR model was assessed with a correlation coeffect of cross-validation (Rcal) and predictions (Rval), standard error of cross-validation (SECV) and predictions (SEP), outlier diagnostics, leverage, and residual analysis [25]. Samples with high residuals and leverage were re-analyzed and excluded from the model if needed.

3. Results and Discussion

3.1. Characterization of Maple Syrup Samples

Reference analysis results for total soluble solids (°Brix), sugar (sucrose, fructose, and glucose), and total phenolics for all samples, including traditional maple syrup, bourbon barrel (BBL)-aged maple syrup, and table syrups (corn, cane, and mixture—consisted of cane, maple, and agave syrups) are summarized in Table 1.

Table 1. Reference analysis results of total soluble solids, sugar (sucrose, fructose, and glucose), and total phenolics in traditional, bourbon barrel (BBL)-aged maple syrup and commercial table syrups.

		Traditional Maple Syrup (n = 19)		BBL Maple Syrup (n = 13)	Table Syrups (n = 5)
°Brix	Minimum	65.51		65.39	39.63
	Maximum	67.65		68.69	78.27
	Mean	66.57		66.56	67.64
	SD	0.55		0.87	14.32
Sucrose (%, g/100 g)	Minimum	22.02		60.13	3.51
	Maximum	67.60		69.42	51.49
	Mean	57.56		63.72	21.75
	SD	14.78		2.73	17.77
Fructose (%, g/100 g)	Minimum	0.00		0.00	12.62
	Maximum	17.14		0.00	14.36
	Mean	2.01		0.00	13.31
	SD	4.86		0.00	0.76
Glucose (%, g/100 g)	Minimum	0.00		0.00	9.75
	Maximum	17.06		0.00	14.11
	Mean	2.31		0.00	12.34
	SD	5.48		0.00	1.86
		Golden and Amber (n = 10)	Dark (n = 5)	BBL Maple Syrup (n= 13)	Table Syrups (n = 5)
Total phenolics (μg GAE/mL) [a]	Minimum	115.64	387.01	317.37	NA [c]
	Maximum	338.94	582.39	713.40	NA
	Mean	271.15	479.53	458.25	NA
	SD	64.93	72.85	124.78	NA
	p-Value		<0.001 [b]		NA

[a] Total phenolics, expressed as micrograms of gallic acid equivalent (GAE) per 1 mL of distilled water. Three unusual maple syrups were excluded from this analysis due to containing of interferences. [b] p value, based on one-way ANOVA test; there were significant differences in total phenolics between three types of products ($p < 0.05$). Based on post hoc LSD, all samples were significantly different, except for BBL and dark maple syrup ($p = 0.69$). [c] Table syrups were excluded from total phenolic analysis.

Traditional maple syrups and BBL maple syrups showed similar total soluble solids (°Brix) contents (Table 1). The °Brix values of maple syrups (65.4–68.7° with an average of 66.6 ± 0.7°) were within the range reported by Stuckel and Low (62.2–74.0° with an average of 67.0 ± 1.6°) [4] and Perkins (66–68°) [26]. Sucrose, fructose, and glucose contents of traditional and BBL maple syrups are summarized in Table 1. We found no significant difference ($p = 0.98$, $p > 0.05$) in the sugar content between traditional and BBL maple syrups. Most sucrose contents of maple syrups agree with the reported literature (51.7–75.6% with an average of 68.0 ± 4.0%) [4], while four labeled as traditional maple syrup samples were far below the range (22.0, 23.5, 36.1, and 50.6%). These same four samples have much higher fructose (9.5–17.1%) and glucose (9.7–17.1%) contents than the literature reports (fructose 0.3 ± 0.5%, and glucose 0.4 ± 1.1%) [4]. In Morselli's study, fructose content in maple syrup was undetectable, and glucose content ranged from 0–7.3% [27]. The glucose to fructose ratio of 3 of the suspect samples was ~1:1, while the other had a 1:1.6 ratio. Invert sugar in maple syrups can be produced from sucrose hydrolysis during thermal processing or microbial contamination of the sap [26]. However, the abnormally high invert sugar contents and low sucrose contents in the samples indicate the potential adulteration of maple syrup with inexpensive table syrups. We evaluated commercial table syrup blends that showed similar levels of fructose (13.3 ± 0.8%), glucose (12.3 ± 1.9%), and sucrose 21.7 ± 7.8%) content to the suspect maple syrups.

Figure 1 shows the representative HPLC chromatograms of traditional maple syrup, suspicious maple syrup (which has high invert sugar and low sucrose content), BBL maple syrup and table syrup (specifically corn syrup). The sugar profiles of both traditional and BBL maple syrups obtained by HPLC showed sucrose as the dominant sugar, while the suspicious maple syrup had noticeably high fructose and glucose contents as well as a detectable but low maltose content. In the literature, it has been stated that authentic maple syrup should not have any detectable maltose content [1]. Furthermore, it has been reported that syrup sweeteners, including molasses, high fructose corn syrup, and honey, have wide maltose composition variability, from 3.0–14.4% [28].

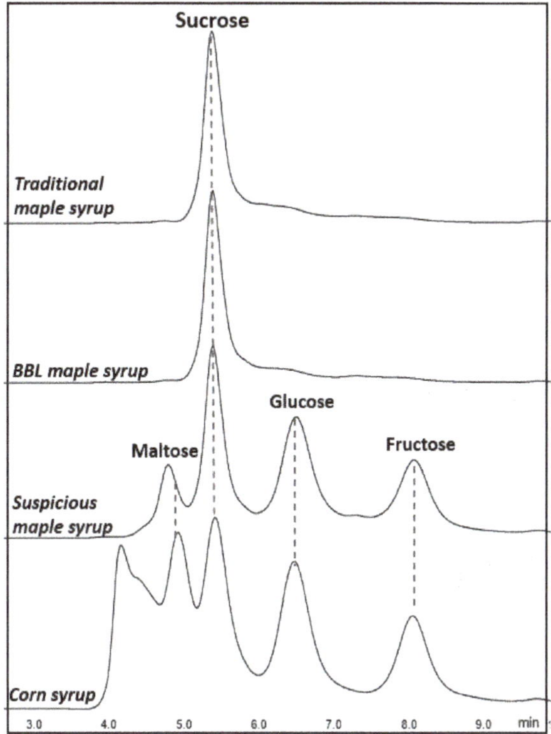

Figure 1. Representative HPLC-RID chromatograms of sugar profiles for traditional maple syrup, BBL maple syrup, suspicious maple syrup and corn syrup.

The total phenolic contents of traditional maple syrups and BBL maple syrups are summarized in Table 1. In previous reports, total phenolic contents in maple syrups ranged from 200–900 µg/mL, which agreed with our findings [29,30]. Since the FC method is based on the reagent's chemical reduction, the most problematic assay interference could be the presence of reducing sugars and samples with high protein levels [19]. In traditional maple syrups (except for suspicious ones) and BBL maple syrups, reducing sugars were undetectable. In addition, according to the literature, protein contents in maple syrups are in low concentration (~0–50 ppm) [1]. Therefore, the FC method can be considered a suitable method for analyzing total phenolics in maple syrups. The suspicious maple syrups were excluded from this analysis due to a high-level of reducing sugar (glucose and fructose) content.

The total phenolic content of traditional maple syrups correlated with their color grade and can be separated into golden and amber and dark groups. The dark traditional and BBL maple syrups had significantly higher total phenolic content than the golden and amber maple syrups ($p < 0.001$) according to one-way ANOVA and post hoc LSD tests. Dark maple saps are collected in the later production season when the temperature is warmer (usually at warm springs) and sucrose is converted to invert sugar due to higher microbial activity [1]. Higher invert sugar contents in maple saps result in a stronger Maillard reaction during sap evaporation, giving darker color and stronger flavor in the final maple syrup products. In addition, a higher cultivation temperature of plants and higher activity of beneficial microbe/pathogen/insect feeding increase the total phenolic compounds, which also explain higher phenolics in dark maple syrups [31]. The higher phenolic content of BBL maple syrups could be associated with the aging process in the barrels, resulting in volatile and non-volatile phenolic compounds from the oak wood being absorbed [6] and contributing to their richer, more complex smell and flavor than traditional pure maple syrups.

The unique volatile profile of all maple syrup samples was characterized by GC-MS analysis, and two out of thirteen BBL-aged maple syrups were flagged as having a different volatile profile than the other BBL-aged maple syrup samples. These two samples did not have a different sugar and total phenolic profile than the other BBL samples. As shown in Table A1, a total of 18 volatile compounds were tentatively identified using the NIST 14.L database through a Mass Spectral Library search and were shown to be present in either maple or liquor products. The representative chromatograms for traditional maple syrup, BBL and suspicious BBL maple syrup are shown in Figure 2. There are several noticeable peak differences between BBL and traditional maple syrups. All authentic BBL maple syrups have one unique peak that other traditional and suspicious BBL (n = 2) maple syrups do not have, which corresponds to 1,1-diethoxy-2-methylpropane. Previous studies found 1,1-diethoxy-2-methylpropane in aged bourbon whiskey [32]. Therefore, authentic BBL maple syrups could absorb this volatile compound from the bourbon residue in barrels during the long aging process. In addition, one of the suspicious BBL had a similar volatile pattern as traditional samples in that they all had significantly lower contents in ethanol, oxalic acid, isoamyl alcohol, furfural and phenylethyl alcohol than authentic BBL samples, indicating that this suspicious BBL sample might have a minimum or no aging process, or the aging barrel does not contain any bourbon residuals [33]. While the other suspicious BBL sample had a similar volatile pattern as traditional samples, except for contents of isoamyl alcohol and oxalic acid, which were even higher than authentic BBLs, indicating that instead of aging, it might be added with bourbon flavor.

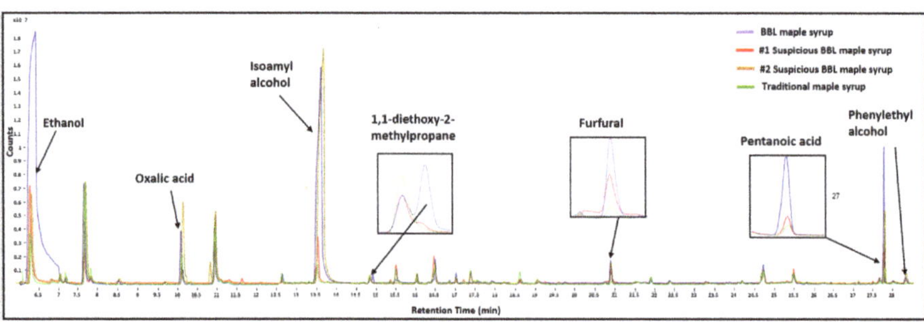

Figure 2. Representative GC-MS chromatograms of volatile compound profiles for BBL maple syrup, suspicious BBL maple syrup and traditional maple syrup.

3.2. Spectral Information of Maple Syrup Samples

The characteristic FT-IR absorption spectra of traditional maple syrup, BBL maple syrup, and corn syrup (as an example of table syrups) and their corresponding band assignments for specific functional groups are shown in Figure 3a. Key absorbance signals included the band at 2929 cm^{-1} associated with C-H stretching of the CH2 group in carbohydrates [8]. The band at 1637 cm^{-1} may be mainly related to O-H bonding in water, with minor contributions to C-O stretching in saccharides [34]. The band at 1415 cm^{-1} related to C-H bending [35] and the band at 1327 cm^{-1} related to O-H bending of the C-OH group might attribute to organic acids. The band at 1110 cm^{-1} was associated with C-O stretching of C-O-C linkage, which could be the glycosidic linkage in sucrose. The bands at 1042 and 990 cm^{-1} were associated with C-O stretching in the C-OH group and C-C stretching in carbohydrates, and the band at 927 cm^{-1} was related to C-H stretching [8]. The broadband located around 3600–3000 cm^{-1} was mainly related to O-H bonds stretching in water, which has been reported previously as the major infrared bands of water located at 3490 and 3280 cm^{-1} for O–H stretching [36,37]. The range from 1200 to 800 cm^{-1} could be assigned to the carbohydrates absorption region, mainly related to sucrose, fructose, and glucose absorption bands [8,36].

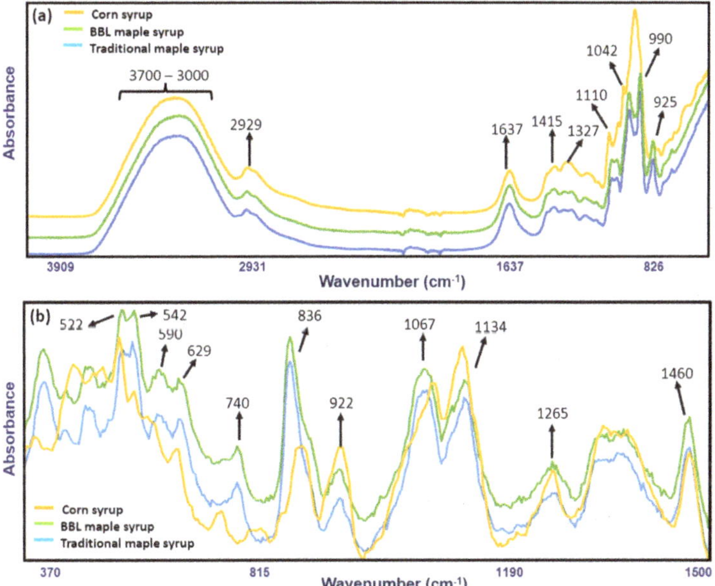

Figure 3. (**a**)FT-IR spectrum band positions and corresponding wavenumbers of traditional maple syrup, BBL maple syrup, and corn syrup at a frequency of 4000–700 cm^{-1} collected using a portable five-reflections ZnSe crystal ATR system. (**b**) Raman spectrum, band positions and corresponding wavenumbers of traditional maple syrup, BBL maple syrup and corn syrup at a frequency of 350–1500 cm^{-1} collected using benchtop Raman with 1064 nm excitation laser.

The characteristic Raman signal of traditional maple syrup, BBL maple syrup, and corn syrup (as an example of table syrups) and their corresponding band assignments for specific functional groups are shown in Figure 3b. The major bands in the Raman spectra were centered in the range of 500–1500 cm^{-1}. One major band at 522 cm^{-1} was associated with the deformation of C-C-O and C-C-C [38], while another major band at 542 cm^{-1} is related to an unassigned vibration [8]. The band at 590 cm^{-1} is associated with skeletal vibration [38], and the band at 629 cm^{-1} corresponded with sugar ring deformation [28]. The minor band at 740 cm^{-1} could be due to C-C, C-O stretching in the carbohydrate molecules [13]. The dominant peak at 835 cm^{-1} is responsible for C-C stretching, which is an intense band found in sucrose [39]. The high Raman signal at 835 cm^{-1} band is associated with the high sucrose content (~68%) in maple syrup [4]. Both peaks at 923 and 1067 cm^{-1} are responsible for the combination of vibration C-H bending, especially the C-H bond at C1 position and COH bending [8]. The peak at 1127 cm^{-1} could be due to the deformation of C-O-H, as well as the vibration of C-N, which is found in protein or amino acid [28,38]. The band at 1265 cm^{-1} is associated with the deformation of C-C-H, O-C-H,

C-O-H, and the vibration of Amide III, which is a peptide bond, and the band at 1460 cm^{-1} is related to the symmetric deformation in the plane of CH2 [38].

In both FT-IR and Raman spectra, corn syrup was easily differentiated from traditional maple syrup and BBL maple syrup using only visual assessment due to maple syrups' unique patterns. However, between traditional maple syrup and BBL maple syrup, the spectral differences were not noticeable via visual evaluation due to their similarity. Therefore, a supervised classification method (SIMCA) was used to analyze the spectral data and to determine the class belongings, including traditional maple syrups, BBL maple syrups, and suspicious samples.

3.3. *Multivariate Data Analysis*

3.3.1. SIMCA Classification Model of GC-MS

The GC-MS data of volatile compounds in traditional and BBL-aged maple syrup samples were analyzed and grouped using the Soft Independence Modeling of Class Analogy (SIMCA), and the class projection plot is shown in Figure A1. All the authentic BBL maple syrups were successfully discriminated from the traditional maple syrups based on their volatile composition, having an interclass distance (ICD) of 4.1. Furthermore, authentic BBL samples were also successfully differentiated from the suspicious BBL maple syrups (ICD = 2.2), and the classification pattern agreed with the GC-MS data that one of the suspicious BBL grouped with traditional samples, while the other one did not fall into either traditional or authentic sample group. Overall, the five most critical volatile compounds that have the highest impact on SIMCA model discrimination are the order of ethanol, isoamyl alcohol, isobutanol, oxalic acid, and acetoin, which are found to exist in bourbon whiskey or maple sap [33,40]. Therefore, these compounds are significant in authenticating qualified BBL maple syrups from suspicious BBL and traditional maple syrups.

3.3.2. SIMCA Classification Models of FT-IR and Raman Spectroscopy

Collected FT-IR and Raman spectra were analyzed using SIMCA classification analysis to discriminate traditional and BBL maple syrups from suspicious maple syrups. The multiple-class approach was applied for both FT-IR and Raman spectral data by having two well-established classes existing (BBL and traditional maple syrups) in the training model. The projection plots of training sets are shown in Figure 4a,c. The training sets were developed using 11 BBL maple syrups (two suspicious BBL samples were excluded) and 15 traditional maple syrups (four suspicious traditional samples were excluded). All the BBL maple syrups were assigned to class number 1, and traditional maple syrups were assigned to a different class (#2). Suspicious maple syrups that were found according to the HPLC and GC-MS analysis were assigned as non-target samples and were not represented by the classes. For the FT-IR model, five factors were employed and explained 99.8% of the variances. In the Raman model, six factors were used and explained 98.1% of the variances. In this approach, the training models have ICDs of 4.8 and 2.5, classifying BBL maple syrups into traditional maple syrups based on the FT-IR and Raman methods, respectively.

The SIMCA discriminating power plot interprets variables that have a predominant effect on the sample classification [41]. The fingerprint region of 800–1200 and 800–1000 cm^{-1} was used to discriminate BBL and traditional maple syrups using FT-IR and Raman spectrometers, respectively. For the FT-IR system, most of the model variance was explained by intensity differences of bands located at 878 cm^{-1}, which is closely related to the symmetric stretching of the primary alcohol group, and 1034 cm^{-1}, which is related to the C-O bond stretching [42,43]. For the Raman system, most of the model variance was explained by the band at 879 cm^{-1}, which was also related to the alcohol group's concentration [44]. Therefore, both FT-IR and Raman methods indicated that differences in compounds with alcoholic groups could explain the variance between BBL and traditional maple syrups. This finding agrees with our GC-MS results since ethanol, isoamyl alcohol, and isobutanol are the top three compounds, assisting BBL maple syrups' differentiation from traditional maple syrups.

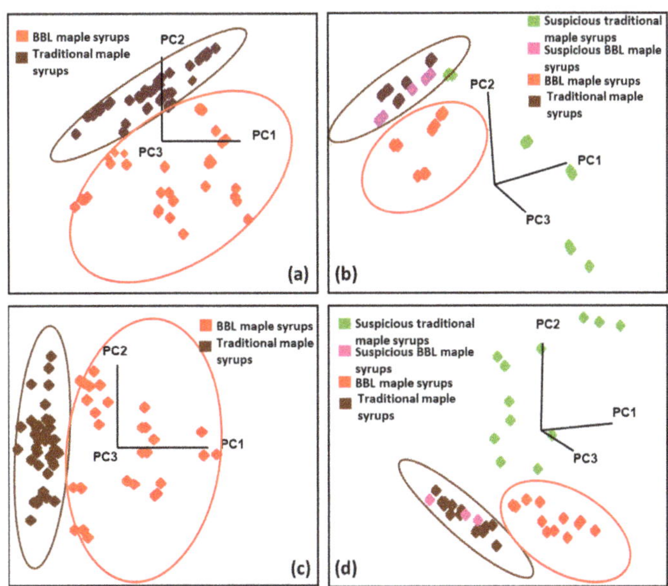

Figure 4. Soft independent modeling of class analogy (SIMCA) projection plots of classification of traditional and BBL maple syrups with (**a**) FT-IR and (**c**) Raman; prediction of external validation sets, including authentic traditional and BBL samples and suspicious samples by (**b**) FT-IR and (**d**) Raman.

The performances of the supervised multiple-class FT-IR and Raman models were evaluated through an independent external validation set, which comprised four traditional and four BBL maple syrups, four suspicious traditional maple syrups, and two suspicious BBL maple syrups. All four traditional and four BBL maple syrups in the external validation set were tested with all reference analyses, and no abnormal pattern was found. The projection plots of validation sets are shown in Figure 4b, d and displayed well-separated clusters in both methods.

Both FT-IR and Raman models accurately predict all traditional and BBL maple syrups in the correct class (n true positive = 8, n false negative = 0, sensitivity = 100%), except for two traditional samples with one replication predicted as No Match in the Raman model. In addition, all suspicious traditional maple syrups were predicted as non-pure, and all suspicious BBL samples were predicted as traditional maple syrup, which was consistent with our expectations (n false positive = 0, n true negative = 6, specificity = 100%). Therefore, all traditional and BBL maple syrups were successfully authenticated by FT-IR and Raman with the multiple-class approach based on their unique chemical composition, and our results agreed with the reference analysis. Our FT-IR and Raman systems displayed a better performance than previous studies of detecting cheap sweetener adulteration in maple syrups, which had 88–100% correctness of discrimination with FT-IR and 98% correctness of discrimination with Raman [8]. Since there is no previous peer-reviewed study investigating BBL maple syrups' characterization and no formal regulation about the quality control of BBL maple syrups, a larger sample size of BBL maple syrup samples is needed for generating a more comprehensive and representative prediction model in the future.

3.3.3. Regression Models

It is important to monitor the °Brix and sucrose contents in maple syrup to ensure product quality and stability [26]. Partial least square regression (PLSR) prediction models were developed with FT-IR and Raman spectra and reference values of °Brix and sucrose contents (Figure A2). Performance statistics of the PLSR models developed using training (n = 26) and external validation (n = 11) data sets are listed in Table 2. The number of samples and the range in training models are not all the same due to outlier exclusion. Four and five factors were selected to generate FT-IR and Raman training models, respectively, according to the standard error of cross-validation (SECV) (leave-out-out) result from carrying out the best quality of the models as well as to avoid possible overfitting.

Table 2. Statistics of partial least square regression (PLSR) models developed using a training (n = 30) and an external validation (n = 7) data set based on FT-IR and Raman spectral data for estimating Brix and sucrose contents in traditional maple syrups, BBL maple syrups, and table syrup samples.

Approach	Sugar	Training Model					External Validation Model			
		Range	N [a]	Factor	SECV [b]	Rcal	Range	N [c]	SEP [d]	Rval
FT-IR	°Brix	39.3–78.7	30	5	0.56	0.99	65.2–78.4	7	0.88	0.98
	Sucrose	3.3–66.2	30	4	1.68	0.99	18.4–65.3	7	1.66	0.99
Raman	°Brix	39.9–78.5	29	5	1.00	0.98	65.0–78.7	7	1.23	0.96
	Sucrose	3.5–66.6	30	3	1.69	0.99	17.5–65.1	7	1.67	0.99

[a] Sample number in the training models. [b] Standard error of cross validation. [c] Sample number in the external validation models. [d] Standard error of prediction.

Our PLSR models showed strong correlations (Rcal > 0.98 and Rval > 0.95) in predicting °Brix and sucrose contents in traditional maple syrups, BBL maple syrups, and table syrup samples. The standard error of prediction (SEP) values were 0.88% and 1.66% for FT-IR validation models for °Brix and sucrose, respectively, and were 1.23% and 1.67% for Raman validation models for °Brix and sucrose, respectively. Similar SECV and SEP were obtained, indicating the robustness of the models. Standard errors of laboratory (SEL) for reference methods of °Brix and sucrose were 0.21% and 0.62%, respectively. The SEL values were compared with the prediction performances of the models (SEP), and we found that the SEP values (Table 2) were always higher than those of SEL because the SEP includes not only the sampling and analysis errors but also the spectroscopy and model errors [45]. The SEP obtained for the FT-IR and Raman models were 2.7 times those of the SEL for sucrose, representing good precision of the models [46]. Conversely, the models predicting °Brix had a SEP/SEL ratio of 4.2 (FT-IR) and 5.9 (Raman), which were higher than the SEP/SEL threshold of 2 [46] for acceptable precision compared to the referenced method. However, our models show superior performance compared to reported °Brix predictions for honey using FTIR (R2val = 0.86, SEP = 1.84%) and Raman (R2val = 0.87, SEP = 1.32%) [47,48]. Nickless et al. quantified the total sugar contents in Manuka nectar using FT-IR, reporting Rval = 0.95 and SEP = 1.17% values [15].

The regression vector plots, shown in Figure A3, help to identify the functional groups whose variance is the highest for correlating between reference values and spectral data. The key FT-IR region for the °Brix and sucrose predictions was in the 1750–700 cm^{-1} range, with distinguished bands centered at 1635 (OH bending vibration characteristic of absorbed water) and the 1125 to 900 cm^{-1} related to C-O stretching and ring vibrational modes of sugars [8,34]. The regression vector plots for Raman data indicated that the bands at 835, 990, 1100 cm^{-1} explained most of the variance for the Brix model, and the bands at 424, 600, and 890 cm^{-1} explained for the Sucrose model. The scattering bands in the vicinity of 424 and 600 cm^{-1} are associated with the deformation of C-C-O and C-C-C [38]. The bands near 990 and 1100 cm^{-1} are related to the deformation modes of saccharides functional groups [28,38].

4. Conclusions

In summary, FT-IR and Raman techniques fingerprinted maple syrup products based on their unique chemical composition, allowing for BBL and traditional maple syrup authentication. Both FT-IR and Raman systems combined with SIMCA provided non-destructive, fast, and accurate determination of quality traits in BBL and traditional maple syrups and detected potential maple syrup adulterants. Our results showed that 15% of commercial maple syrup (traditional and/or BBL) samples that were tested and labeled as "pure" exhibited unusual sugar and/or volatile profiles, and both FT-IR and Raman equipment discriminated these suspicious samples from the pure ones. Furthermore, both FT-IR and Raman, combined with PLSR, showed good predictions for all samples' total °Brix and sucrose contents.

Author Contributions: K.Z.: methodology, validation, formal analysis, investigation, writing—original draft, visualization; D.P.A.: methodology, validation, writing—review and editing, supervision, project administration; L.E.R.-S.: conceptualization, validation, resources, writing—review and editing, supervision. All authors have read and agreed to the published version of the manuscript.

Funding: This research did not receive any specific grant from funding agencies in the public, commercial, or not-for-profit sectors.

Institutional Review Board Statement: Not applicable.

Informed Consent Statement: Not applicable.

Data Availability Statement: The data used to support the findings of this study can be made available by the corresponding author upon request.

Acknowledgments: We would like to thank Bissell Maple Farm, Jefferson, Ohio, for kindly providing maple syrup and BBL samples for our research.

Conflicts of Interest: The authors declare no conflict of interest.

Appendix A

Table A1. Summarization of 18 tentatively identified volatile compounds from maple syrup samples utilizing GC-MS.

		2-Methylpropan-1-Ol	Ethanol	3-Hydroxy-butan-2-One	1,1-Diethoxy-2-Methylpropane	Pentan-1-ol	2-O-Butyl 1-O-Propyl Oxalate	3-Methylbut-3-en-1-oll	Unknown Compound	2-(2-Ethylhexoxy)Ethanol	4-Butoxybutan-2-One	2-Methylcyclopent-2-En-1-One	1-O-(2-Methylpropyl) 4-O-Propan-2-Yl 2,2-Dimethyl-3-Propan-2-Ylbutanedioate	2-Phenylethanol	2-Methylpyrazine	Furan-2-Carbaldehyde	Benzaldehyde	2,6-Dimethyl pyrazine	4-Methylbenzaldehyde
	Ions (m/z)	33-43-74	31-45	28-45-88	47-55-103	41-55-70	33-43-74	41-56-68	57-69-89	45-57-71	28-43-73	57-67-96	43-71-159	65-91-122	77-94-105	39-67-96	51-77-106	28-42-108	65-91-119
BBL	min	7.60×10^5	3.10×10^7	2.70×10^4	1.40×10^5	8.10×10^6	5.20×10^4	1.30×10^4	8.70×10^2	2.10×10^5	1.10×10^3	9.20×10^3	1.40×10^5	1.00×10^5	2.40×10^4	5.70×10^5	5.00×10^4	1.40×10^4	4.80×10^3
(n = 11)	max	6.10×10^6	8.80×10^7	3.80×10^5	2.70×10^6	3.10×10^7	6.10×10^6	1.60×10^5	3.70×10^5	4.90×10^5	3.20×10^5	1.90×10^5	7.70×10^6	7.20×10^5	2.40×10^5	2.70×10^6	2.10×10^5	1.90×10^5	6.80×10^5
	mean	3.30×10^6	6.10×10^7	1.40×10^5	5.90×10^5	1.90×10^7	3.20×10^6	7.10×10^4	1.40×10^5	3.30×10^5	1.40×10^5	6.60×10^4	4.50×10^6	2.60×10^5	7.60×10^4	1.40×10^6	1.00×10^5	7.20×10^4	3.30×10^5
Abnormal BBL	min	7.60×10^5	8.90×10^6	5.20×10^5	0.00	2.10×10^6	7.60×10^5	1.60×10^5	1.80×10^5	4.20×10^5	6.90×10^4	5.30×10^4	1.80×10^5	2.30×10^5	1.40×10^5	8.00×10^5	7.90×10^4	1.40×10^5	4.10×10^5
(n = 2)	max	6.00×10^6	9.40×10^6	8.90×10^5	0.00	3.20×10^7	6.00×10^6	2.70×10^5	2.90×10^5	5.20×10^5	2.20×10^5	2.20×10^5	2.50×10^5	6.00×10^5	2.10×10^6	1.70×10^6	8.80×10^4	1.20×10^6	5.30×10^5
	mean	3.40×10^6	9.10×10^6	7.10×10^5	0.00	1.70×10^7	3.40×10^6	2.10×10^5	2.40×10^5	4.70×10^5	1.40×10^5	1.40×10^5	2.20×10^5	4.20×10^5	1.10×10^6	1.20×10^6	8.40×10^4	6.70×10^5	4.70×10^5
Golden and Amber	min	2.70×10^4	2.90×10^5	6.90×10^2	0.00	9.10×10^3	9.70×10^3	2.60×10^3	5.20×10^3	6.10×10^4	1.30×10^3	4.00×10^3	2.90×10^2	3.20×10^3	8.10×10^2	1.90×10^3	8.10×10^2	1.20×10^3	2.30×10^2
(n = 10)	max	1.10×10^7	9.40×10^6	9.70×10^5	0.00	9.80×10^5	2.50×10^6	5.00×10^5	5.80×10^5	7.70×10^5	3.60×10^5	1.50×10^5	3.40×10^5	2.50×10^4	3.70×10^5	5.00×10^5	1.30×10^5	3.80×10^5	8.10×10^5
	mean	2.40×10^6	2.30×10^6	4.90×10^5	0.00	3.30×10^5	8.90×10^5	2.10×10^5	1.90×10^5	3.70×10^5	1.60×10^5	6.30×10^4	1.60×10^5	1.60×10^4	8.80×10^4	1.50×10^5	6.40×10^4	1.10×10^5	3.50×10^5
Dark	min	9.90×10^5	4.30×10^6	6.90×10^5	0.00	2.90×10^5	9.90×10^5	1.80×10^5	1.60×10^5	3.30×10^5	8.20×10^4	6.70×10^4	9.80×10^4	3.00×10^4	2.00×10^5	2.50×10^5	5.80×10^4	4.00×10^5	2.60×10^5
(n = 5)	max	2.70×10^6	1.70×10^7	1.00×10^6	0.00	3.00×10^6	2.70×10^6	3.00×10^5	2.20×10^5	5.30×10^5	2.90×10^5	5.30×10^5	2.30×10^5	6.50×10^4	1.60×10^6	5.20×10^5	1.40×10^5	1.30×10^6	4.40×10^5
	mean	1.90×10^6	1.00×10^7	8.80×10^5	0.00	1.50×10^6	1.90×10^6	2.60×10^5	2.00×10^5	4.50×10^5	2.00×10^5	2.20×10^5	1.60×10^5	4.90×10^4	7.00×10^5	3.40×10^5	8.40×10^4	6.40×10^5	3.80×10^5
	p-value	<0.001	<0.001	<0.001	<0.001	<0.001	0.002	0.009	0.705	0.403	0.755	0.15	<0.001	<0.001	<0.001	<0.001	0.163	<0.001	0.932
	Ions (m/z)	33-43-74	31-45	28-45-88	47-55-103	41-55-70	33-43-74	41-56-68	57-69-89	45-57-71	28-43-73	57-67-96	43-71-159	65-91-122	77-94-105	39-67-96	51-77-106	28-42-108	65-91-119
BBL	min	7.60×10^5	3.10×10^7	2.70×10^4	1.40×10^5	8.10×10^6	5.20×10^4	1.30×10^4	8.70×10^2	2.10×10^5	1.10×10^3	9.20×10^3	1.40×10^5	1.00×10^5	2.40×10^4	5.70×10^5	5.00×10^4	1.40×10^4	4.80×10^3
(n = 11)	max	6.10×10^6	8.80×10^7	3.80×10^5	2.70×10^6	3.10×10^7	6.10×10^6	1.60×10^5	3.70×10^5	4.90×10^5	3.20×10^5	1.90×10^5	7.70×10^6	7.20×10^5	2.40×10^5	2.70×10^6	2.10×10^5	1.90×10^5	6.80×10^5
	mean	3.30×10^6	6.10×10^7	1.40×10^5	5.90×10^5	1.90×10^7	3.20×10^6	7.10×10^4	1.40×10^5	3.30×10^5	1.40×10^5	6.60×10^4	4.50×10^6	2.60×10^5	7.60×10^4	1.40×10^6	1.00×10^5	7.20×10^4	3.30×10^5
Abnormal BBL	min	7.60×10^5	8.90×10^6	5.20×10^5	0.00	2.10×10^6	7.60×10^5	1.60×10^5	1.80×10^5	4.20×10^5	6.90×10^4	5.30×10^4	1.80×10^5	2.30×10^5	1.40×10^5	8.00×10^5	7.90×10^4	1.40×10^5	4.10×10^5
(n = 2)	max	6.00×10^6	9.40×10^6	8.90×10^5	0.00	3.20×10^7	6.00×10^6	2.70×10^5	2.90×10^5	5.20×10^5	2.20×10^5	2.20×10^5	2.50×10^5	6.00×10^5	2.10×10^6	1.70×10^6	8.80×10^4	1.20×10^6	5.30×10^5
	mean	3.40×10^6	9.10×10^6	7.10×10^5	0.00	1.70×10^7	3.40×10^6	2.10×10^5	2.40×10^5	4.70×10^5	1.40×10^5	1.40×10^5	2.20×10^5	4.20×10^5	1.10×10^6	1.20×10^6	8.40×10^4	6.70×10^5	4.70×10^5
Golden and Amber	min	2.70×10^4	2.90×10^5	6.90×10^2	0.00	9.10×10^3	9.70×10^3	2.60×10^3	5.20×10^3	6.10×10^4	1.30×10^3	4.00×10^3	2.90×10^2	3.20×10^3	8.10×10^2	1.90×10^3	8.10×10^2	1.20×10^3	2.30×10^2
(n = 10)	max	1.10×10^7	9.40×10^6	9.70×10^5	0.00	9.80×10^5	2.50×10^6	5.00×10^5	5.80×10^5	7.70×10^5	3.60×10^5	1.50×10^5	3.40×10^5	2.50×10^4	3.70×10^5	5.00×10^5	1.30×10^5	3.80×10^5	8.10×10^5
	mean	2.40×10^6	2.30×10^6	4.90×10^5	0.00	3.30×10^5	8.90×10^5	2.10×10^5	1.90×10^5	3.70×10^5	1.60×10^5	6.30×10^4	1.60×10^5	1.60×10^4	8.80×10^4	1.50×10^5	6.40×10^4	1.10×10^5	3.50×10^5
Dark	min	9.90×10^5	4.30×10^6	6.90×10^5	0.00	2.90×10^5	9.90×10^5	1.80×10^5	1.60×10^5	3.30×10^5	8.20×10^4	6.70×10^4	9.80×10^4	3.00×10^4	2.00×10^5	2.50×10^5	5.80×10^4	4.00×10^5	2.60×10^5
(n = 5)	max	2.70×10^6	1.70×10^7	1.00×10^6	0.00	3.00×10^6	2.70×10^6	3.00×10^5	2.20×10^5	5.30×10^5	2.90×10^5	5.30×10^5	2.30×10^5	6.50×10^4	1.60×10^6	5.20×10^5	1.40×10^5	1.30×10^6	4.40×10^5
	mean	1.90×10^6	1.00×10^7	8.80×10^5	0.00	1.50×10^6	1.90×10^6	2.60×10^5	2.00×10^5	4.50×10^5	2.00×10^5	2.20×10^5	1.60×10^5	4.90×10^4	7.00×10^5	3.40×10^5	8.40×10^4	6.40×10^5	3.80×10^5
	p-value	<0.001	<0.001	<0.001	<0.001	<0.001	0.002	0.009	0.705	0.403	0.755	0.15	<0.001	<0.001	<0.001	<0.001	0.163	<0.001	0.932

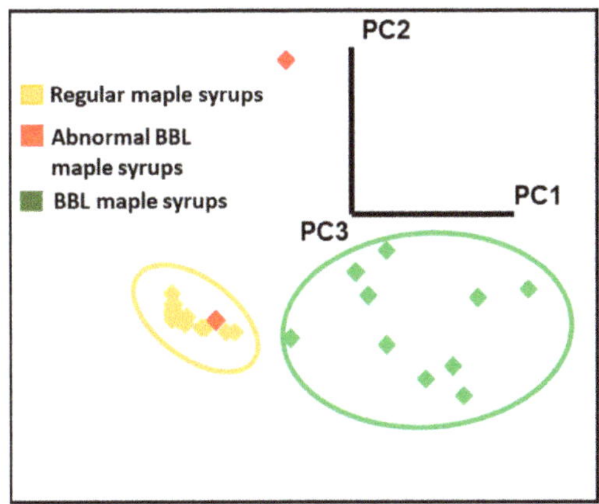

Figure A1. Soft independent modeling of class analogy (SIMCA) projection plots of classification of authentic BBL samples from suspicious BBL and traditional maple syrup samples by GC-MS.

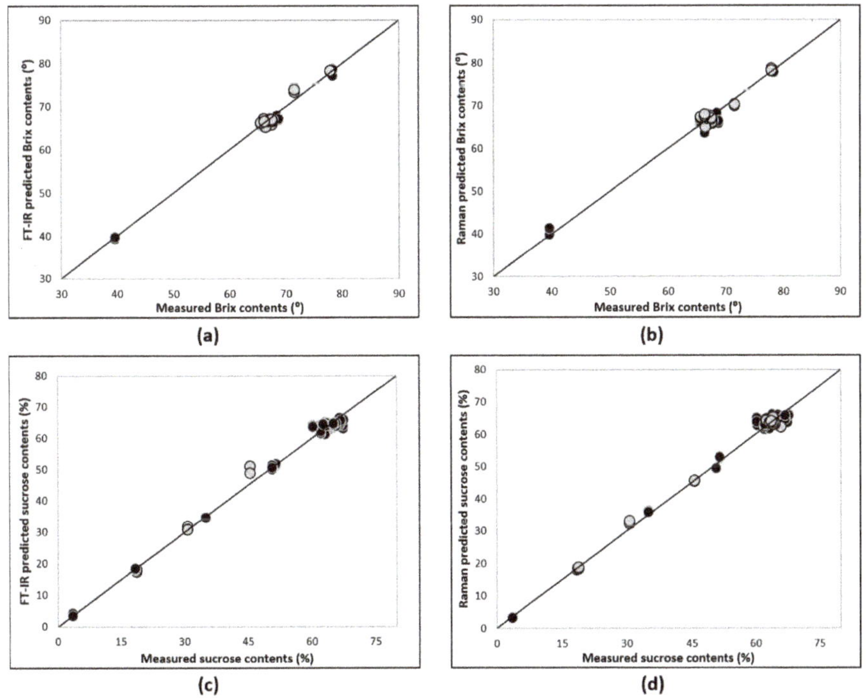

Figure A2. PLSR calibration and validation plots for Brix (**a**,**b**), and sucrose (**c**,**d**) in traditional maple syrups, BBL maple syrups, and table syrup samples utilizing 4500 FT-IR and Raman data, respectively. Black circles represent calibration set samples; gray circles represent external validation set samples.

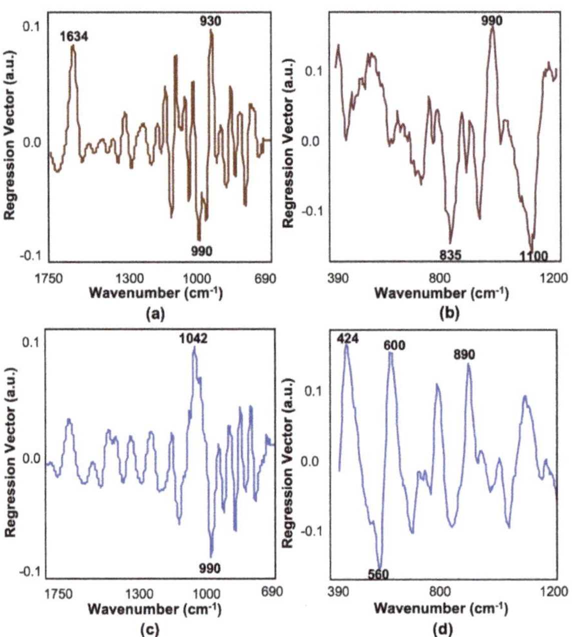

Figure A3. PLSR regression vectors for Brix (**a**,**b**) and sucrose (**c**,**d**), utilizing 4500 FT−IR and Raman data, respectively.

References

1. Heiligmann, R.B.; Koelling, M.R.; Perkins, T.D. *North American Maple Syrup Producers Manual*; Ohio State University Extension: Columbus, OH, USA, 2006; p. 29.
2. Chamberlain, A.F. The Maple amongst the Algonkian Tribes. *Am. Anthropol.* **1891**, *4*, 39–44. [CrossRef]
3. USDA Wisconsin Ag News—Maple Syrup. Available online: https://www.nass.usda.gov/Statistics_by_State/Wisconsin/Publications/Crops/2020/WI-Maple-Syrup-06-20.pdf (accessed on 10 September 2021).
4. Stuckel, J.G.; Low, N.H. The chemical composition of 80 pure maple syrup samples produced in North America. *Food Res. Int.* **1996**, *29*, 373–379. [CrossRef]
5. St-Pierre, P.; Pilon, G.; Dumais, V.; Dion, C.; Dubois, M.J.; Dubé, P.; Desjardins, Y.; Marette, A. Comparative analysis of maple syrup to other natural sweeteners and evaluation of their metabolic responses in healthy rats. *J. Funct. Foods* **2014**, *11*, 460–471. [CrossRef]
6. Li, S.; Duan, C. Astringency, bitterness and color changes in dry red wines before and during oak barrel aging: An updated phenolic perspective review. *Crit. Rev. Food Sci. Nutr.* **2019**, *59*, 1840–1867. [CrossRef]
7. Samuelson, E. What Is Bourbon Barrel Aged Maple Syrup? Available online: https://www.eatlikenoone.com/bourbon-barrel-aged-maple-syrup-2.htm (accessed on 10 September 2021).
8. Paradkar, M.M.; Sivakesava, S.; Irudayaraj, J. Discrimination and classification of adulterants in maple syrup with the use of infrared spectroscopic techniques. *J. Sci. Food Agric.* **2002**, *82*, 497–504. [CrossRef]
9. Carro, O.; Hillaire-Marcel, C.; Gagnon, M. Detection of Adulterated Maple Products by Stable Carbon Isotope Ratio. *J. AOAC Int.* **1980**, *63*, 840–844. [CrossRef]
10. Stuckel, J.G.; Low, N.H. Maple Syrup Authenticity Analysis by Anion-Exchange Liquid Chromatography with Pulsed Amperometric Detection. *J. Agric. Food Chem.* **1995**, *43*, 3046–3051. [CrossRef]
11. Tremblay, P.; Paquin, R. Improved Detection of Sugar Addition to Maple Syrup Using Malic Acid as Internal Standard and in 13 C Isotope Ratio Mass Spectrometry (IRMS). *J. Agric. Food Chem.* **2007**, *55*, 197–203. [CrossRef]
12. Santos, P.M.; Pereira-Filho, E.R.; Rodriguez-Saona, L.E. Application of hand-held and portable infrared spectrometers in bovine milk analysis. *J. Agric. Food Chem.* **2013**, *61*, 1205–1211. [CrossRef]
13. Paradkar, M.M.; Sakhamuri, S.; Irudayaraj, J. Comparison of FTIR, FT-Raman, and NIR Spectroscopy in a Maple Syrup Adulteration Study. *J. Food Sci.* **2002**, *67*, 2009–2015. [CrossRef]
14. Lu, X.; Al-Qadiri, H.M.; Lin, M.; Rasco, B.A. Application of Mid-infrared and Raman Spectroscopy to the Study of Bacteria. *Food Bioprocess Technol.* **2011**, *4*, 919–935. [CrossRef]

15. Nickless, E.M.; Holroyd, S.E.; Hamilton, G.; Gordon, K.C.; Wargent, J.J. Analytical method development using FTIR-ATR and FT-Raman spectroscopy to assay fructose, sucrose, glucose and dihydroxyacetone, in *Leptospermum scoparium* nectar. *Vib. Spectrosc.* **2016**, *84*, 38–43. [CrossRef]
16. Mellado-Mojica, E.; Seeram, N.P.; López, M.G. Comparative analysis of maple syrups and natural sweeteners: Carbohydrates composition and classification (differentiation) by HPAEC-PAD and FTIR spectroscopy-chemometrics. *J. Food Compos. Anal.* **2016**, *52*, 1–8. [CrossRef]
17. Lachenmeier, D.W. Rapid quality control of spirit drinks and beer using multivariate data analysis of Fourier transform infrared spectra. *Food Chem.* **2007**, *101*, 825–832. [CrossRef]
18. Karoui, R.; Debaerdemaeker, J. A review of the analytical methods coupled with chemometric tools for the determination of the quality and identity of dairy products. *Food Chem.* **2007**, *102*, 621–640. [CrossRef]
19. Waterhouse, A.L. Determination of Total Phenolics. In *Current Protocols in Food Analytical Chemistry*; John Wiley & Sons, Inc.: Hoboken, NJ, USA, 2003.
20. Kovalenko, I.V.; Rippke, G.R.; Hurburgh, C.R. Determination of amino acid composition of soybeans (*Glycine max*) by near-infrared spectroscopy. *J. Agric. Food Chem.* **2006**, *54*, 3485–3491. [CrossRef]
21. Iacobucci, D.; Schneider, M.J.; Popovich, D.L.; Bakamitsos, G.A. Mean centering helps alleviate "micro" but not "macro" multicollinearity. *Behav. Res. Methods* **2016**, *48*, 1308–1317. [CrossRef]
22. Zimmermann, B.; Kohler, A. Optimizing savitzky-golay parameters for improving spectral resolution and quantification in infrared spectroscopy. *Appl. Spectrosc.* **2013**, *67*, 892–902. [CrossRef]
23. Cozzolino, D. The sample, the spectra and the maths-The critical pillars in the development of robust and sound applications of vibrational spectroscopy. *Molecules* **2020**, *25*, 3674. [CrossRef]
24. Massart, D.; Vandeginste, B.; Deming, S.; Michotte, Y.; Kaufman, L. *Data Handling in Science and Technology*; Elsevier Ltd.: Amsterdam, The Netherlands, 2001; ISBN 9780444828538.
25. Siegmann, B.; Jarmer, T. Comparison of different regression models and validation techniques for the assessment of wheat leaf area index from hyperspectral data. *Int. J. Remote Sens.* **2015**, *36*, 4519–4534. [CrossRef]
26. Perkins, T.D.; van den Berg, A.K. *Chapter 4 Maple Syrup-Production, Composition, Chemistry, and Sensory Characteristics*, 1st ed.; Elsevier Inc.: Amsterdam, The Netherlands, 2009; Volume 56, ISBN 9780123744395.
27. Morselli, M.F. Nutrition Value of Pure Maple Syrup. *Maple Syrup Dig.* **1975**, *14*, 12.
28. Aykas, D.P.; Shotts, M.; Rodriguez-Saona, L.E. Authentication of commercial honeys based on Raman fingerprinting and pattern recognition analysis. *Food Control* **2020**, *117*, 107346. [CrossRef]
29. Singh, A.S.; Jones, A.M.P.; Saxena, P.K. Variation and Correlation of Properties in Different Grades of Maple Syrup. *Plant Foods Hum. Nutr.* **2014**, *69*, 50–56. [CrossRef] [PubMed]
30. Apostolidis, E.; Li, L.; Lee, C.; Seeram, N.P. In vitro evaluation of phenolic-enriched maple syrup extracts for inhibition of carbohydrate hydrolyzing enzymes relevant to type 2 diabetes management. *J. Funct. Foods* **2011**, *3*, 100–106. [CrossRef]
31. Wallis, C.M.; Galarneau, E.R.-A. Phenolic Compound Induction in Plant-Microbe and Plant-Insect Interactions: A Meta-Analysis. *Front. Plant Sci.* **2020**, *11*, 580753. [CrossRef]
32. Aoshima, H.; Hossain, S.; Koda, H.; Kiso, Y. Why Is an Aged Whiskey Highly Valued? *Curr. Nutr. Food Sci.* **2009**, *5*, 204–208. [CrossRef]
33. Ball, D.W. The chemical composition of maple syrup. *J. Chem. Educ.* **2007**, *84*, 1647–1650. [CrossRef]
34. Li, J.R.; Sun, S.Q.; Wang, X.X.; Xu, C.H.; Chen, J.B.; Zhou, Q.; Lu, G.H. Differentiation of five species of *Danggui* raw materials by FTIR combined with 2D-COS IR. *J. Mol. Struct.* **2014**, *1069*, 229–235. [CrossRef]
35. Sekkal, M.; Dincq, V.; Legrand, P.; Huvenne, J.P. Investigation of the glycosidic linkages in several oligosaccharides using FT-IR and FT Raman spectroscopies. *J. Mol. Struct.* **1995**, *349*, 349–352. [CrossRef]
36. Sinelli, N.; Spinardi, A.; Di Egidio, V.; Mignani, I.; Casiraghi, E. Evaluation of quality and nutraceutical content of blueberries (*Vaccinium corymbosum* L.) by near and mid-infrared spectroscopy. *Postharvest Biol. Technol.* **2008**, *50*, 31–36. [CrossRef]
37. Gok, S.; Severcan, M.; Goormaghtigh, E.; Kandemir, I.; Severcan, F. Differentiation of Anatolian honey samples from different botanical origins by ATR-FTIR spectroscopy using multivariate analysis. *Food Chem.* **2015**, *170*, 234–240. [CrossRef]
38. Fernández-Pierna, J.A.; Abbas, O.; Dardenne, P.; Baeten, V. Discrimination of Corsican honey by FT-Raman spectroscopy and chemometrics. *Biotechnol. Agron. Soc. Environ.* **2011**, *15*, 75–84.
39. Salvador, L.; Guijarro, M.; Rubio, D.; Aucatoma, B.; Guillén, T.; Jentzsch, P.V.; Ciobotă, V.; Stolker, L.; Ulic, S.; Vásquez, L.; et al. Exploratory monitoring of the quality and authenticity of commercial honey in Ecuador. *Foods* **2019**, *8*, 105. [CrossRef]
40. Lehtonen, P.J.; Keller, L.D.A.; Ali-Mattila, E.T. Multi-method analysis of matured distilled alcoholic beverages for brand identification. *Eur. Food Res. Technol.* **1999**, *208*, 413–417. [CrossRef]
41. Suhandy, D.; Yulia, M. The use of UV spectroscopy and SIMCA for the authentication of Indonesian honeys according to botanical, entomological and geographical origins. *Molecules* **2021**, *26*, 915. [CrossRef]
42. Picque, D.; Lieben, P.; Corrieu, G.; Cantagrel, R.; Lablanquie, O.; Snakkers, G. Discrimination of Cognacs and other distilled drinks by mid-infrared spectropscopy. *J. Agric. Food Chem.* **2006**, *54*, 5220–5226. [CrossRef]
43. Sharma, K.; Sharma, S.P.; Lahiri, S. Novel Method for Identification and Quantification of Methanol and Ethanol in Alcoholic Beverages by Gas Chromatography-Fourier Transform Infrared Spectroscopy and Horizontal Attenuated Total Reflectance Fourier Transform Infrared Spectroscopy. *J. AOAC Int.* **2009**, *92*, 518–526. [CrossRef]

44. Sivakesava, S.; Irudayaraj, J.; Demirci, A. Monitoring a bioprocess for ethanol production using FT-MIR and FT-Raman spectroscopy. *J. Ind. Microbiol. Biotechnol.* **2001**, *26*, 185–190. [CrossRef]
45. Grassi, S.; Jolayemi, O.S.; Giovenzana, V.; Tugnolo, A.; Squeo, G.; Conte, P.; De Bruno, A.; Flamminii, F.; Casiraghi, E.; Alamprese, C. Near Infrared Spectroscopy as a Green Technology for the Quality Prediction of Intact Olives. *Foods* **2021**, *10*, 1042. [CrossRef]
46. Shenk, J.S.; Westerhaus, M.O. Calibration the ISI Way. In *Near Infrared Spectroscopy: The Future Waves*; Davis, A.M.C., Williams, P., Eds.; NIR Publications: Chichester, UK, 1996.
47. Anguebes, F.; Pat, L.; Ali, B.; Guerrero, A.; Córdova, A.V.; Abatal, M.; Garduza, J.P. Application of Multivariable Analysis and FTIR-ATR Spectroscopy to the Prediction of Properties in Campeche Honey. *J. Anal. Methods Chem.* **2016**, *2016*, 5427526. [CrossRef]
48. Anguebes-Franseschi, F.; Abatal, M.; Pat, L.; Flores, A.; Quiroz, A.V.C.; Ramírez-Elias, M.A.; Pedro, L.S.; Tzuc, O.M.; Bassam, A. Raman Spectroscopy and Chemometric Modeling to Predict Physical-Chemical Honey Properties from Campeche, Mexico. *Molecules* **2019**, *24*, 4091. [CrossRef]

Article

Comparison of Spectroscopy-Based Methods and Chemometrics to Confirm Classification of Specialty Coffees

Verônica Belchior [1,2], Bruno G. Botelho [3] and Adriana S. Franca [2,4,*]

1 The Coffee Sensorium Project, Av. Getulio Vargas, 159, Dores de Campos 36213-000, MG, Brazil; belchior.veronica@gmail.com
2 PPGCA, Universidade Federal de Minas Gerais, Av. Antônio Carlos, 6627, Belo Horizonte 31270-901, MG, Brazil
3 DQ, Universidade Federal de Minas Gerais, Av. Antônio Carlos, 6627, Belo Horizonte 31270-901, MG, Brazil; botelhobrunog@gmail.com
4 DEMEC, Universidade Federal de Minas Gerais, Av. Antônio Carlos, 6627, Belo Horizonte 31270-901, MG, Brazil
* Correspondence: adriana@demec.ufmg.br

Abstract: The Specialty Coffee Association (SCA) sensory analysis protocol is the methodology that is used to classify specialty coffees. However, because the sensory analysis is sensitive to the taster's training, cognitive psychology, and physiology, among other parameters, the feasibility of instrumental approaches has been recently studied for complementing such analyses. Spectroscopic methods, mainly near infrared (NIR) and mid infrared (FTIR—Fourier Transform Infrared), have been extensively employed for food quality authentication. In view of the aforementioned, we compared NIR and FTIR to distinguish different qualities and sensory characteristics of specialty coffee samples in the present study. Twenty-eight green coffee beans samples were roasted (in duplicate), with roasting conditions following the SCA protocol for sensory analysis. FTIR and NIR were used to analyze the ground and roasted coffee samples, and the data then submitted to statistical analysis to build up PLS models in order to confirm the quality classifications. The PLS models provided good predictability and classification of the samples. The models were able to accurately predict the scores of specialty coffees. In addition, the NIR spectra provided relevant information on chemical bonds that define specialty coffee in association with sensory aspects, such as the cleanliness of the beverage.

Keywords: FTIR; NIRS; specialty coffee; PLS models

Citation: Belchior, V.; Botelho, B.G.; Franca, A.S. Comparison of Spectroscopy-Based Methods and Chemometrics to Confirm Classification of Specialty Coffees. *Foods* **2022**, *11*, 1655. https://doi.org/10.3390/foods11111655

Academic Editor: Daniel Cozzolino

Received: 5 May 2022
Accepted: 2 June 2022
Published: 4 June 2022

Publisher's Note: MDPI stays neutral with regard to jurisdictional claims in published maps and institutional affiliations.

1. Introduction

Brazil is the world's largest coffee producer. The recent export data report that Brazil has shipped around 22.872 million bags (60 kg each) from July 2021 to January 2022. Specialty coffees accounted for 17.4% of total Brazilian exports, with an average price of USD 292.44 per bag, representing 23.4% of the total obtained with the shipments in January 2022 [1]. Specialty coffees, defined as high-quality products, are quite relevant for the coffee industry given the higher prices attained in comparison to commodity coffees. While a regular bag of regular green coffee costs approximately USD 200, specialty coffees can go up to USD 1000 per bag.

The quality of a cup of coffee begins in the field. Several factors including coffee species and variety, harvesting, post-harvesting conditions, blend elaboration, and roasting parameters, have a significant influence on the flavor and aroma of the drink. The delicate taste and aroma obtained from a cup of specialty coffee results from a complex combination of physical transformations and chemical reactions that start on the seed and end on the beverage preparation [2,3].

The most common way to evaluate the quality of a green coffee is by cup tasting [3,4]. Several industries, including perfume, coffee and tea, wine, beer, and tobacco, often employ trained personnel for sensory evaluation. In the specific case of coffee, such people are called "Q-graders" and trained to define the sensory profile of different samples. Then, according to the SCA (Specialty Coffee Association) protocol to evaluate coffee, they classify samples by giving different scores [5].

The SCA protocols are based on objective assessment methods, including the presence or absence of sweetness and defects, thus minimizing subjectivity compared to other methodologies. In addition, Q-graders are considered excellent and accurate in giving the scores related to quality, although some errors and inconsistencies regarding the description of a coffee are reported [6].

Furthermore, sensory analysis can lead to a few problems. Bias that comes from the preference and previous knowledge of a specific sample, as well as the influence of some external factors [6] can affect the analysis. Additionally, the Q-grader´s health during the cupping as well as modification on his (her) personal evaluation abilities over time can also affect the results. Such issues can be minimized by using alternative evaluation tools in order to make the coffee trading market more reliable [3,7]. Sensory analysis can also be viewed as a sensitive and time-consuming technique, given the need for well-trained personnel. Considering the economic relevance of specialty coffees in the world trade market, finding alternative tools to confirm coffee quality is of utmost importance.

Many studies have shown the potential of spectroscopic methods in food analysis, with near (NIR) and mid (FTIR) infrared among the most used methods [8]. The employment of such techniques for coffee analysis has been widely reported [9]. Applications include discrimination between coffee species and varieties [10], adulteration of roasted and ground coffee [11–13], and identification of low quality (defective) coffee beans [14–16]. Given that such low quality coffees have a significant effect on the sensory profile of the beverage, spectroscopic methods can also be used to detect differences in sensory parameters. In recent studies, our research group employed chemometrics to develop models for the classification and discrimination between espresso coffee beverages based on generic parameters (intensity and a few sensory aspects) informed by the manufacturers, and also based on sensory analysis performed by a trained panel [7,17]. It was also possible to develop models that classified coffees by cup quality parameters based on classification criteria that are specific to Brazil [4]. Some recent results from another group also showed the feasibility of mid-infrared and chemometrics to discriminate specialty coffees with different roasting profiles [18]. Our latest study showed that FTIR can be successfully used to discriminate specialty coffees classified by Q-graders [3], with models capable of predicting classification scores with high accuracy (validation coefficients above 0.97). Published studies confirm that both FTIR and NIR are promising techniques for coffee quality evaluation. However, in the case of NIR, only qualitative discrimination was performed with respect to coffee quality parameters given by Q-graders, without any attempt to provide an actual score-based classification. Furthermore, a comparison of spectroscopy-based techniques to evaluate specialty coffees has not yet been reported. Since both FTIR and NIR have been shown as reliable techniques for coffee quality definition, a comparison of these methods can indicate which method is more reliable. Although several studies have been described and tested with both techniques, there is still a need for further investigation, in order to improve the quality of predictive models to be applied for food quality evaluation [19].

Therefore, in this study, the potential of NIR was evaluated for establishing sensory characteristics of specialty coffees in terms of quantitative scores. Partial Least Squares (PLS) Regression was employed to build models in order to predict and establish a SCA-based sensory profile. NIR-based models were compared to FTIR ones that were developed in a previous study [3]. To the best of our knowledge, this is the first study in the literature that addresses such comparison for specialty coffee quality evaluation. Furthermore, this is the first work showing that NIR can provide quantitative quality scores.

2. Methodology

2.1. Roasting Tests and Sensory Evaluation

Arabica coffee samples submitted to dry (natural coffee) and wet (pulped natural coffee) processing were employed in the present study. Detailed information regarding sample provenance and quality scores (ranging from 81 to 91) is presented as Supplementary Materials (Table S1) and discussed in our previous study on FTIR analysis of specialty coffees [3]. A summarized description of sample preparation is presented as follows. The samples were roasted in accordance with the SCA protocol for coffee sensory analysis, using an IKAWA® Sample Roaster Pro (London, UK). Individual samples consisted of 50 g of green coffee that were submitted to roasting at temperatures ranging from 170 °C to 227 °C. The roasting time was 4 min 34 s. Roasting tests were performed in duplicate. A total of 56 samples were obtained. These samples were ground using a Porlex Mini® grinder (Porlex Grinders, Osaka, Japan) in order to obtain a fine and homogeneous grind (particle diameter below 0.150 mm). The samples were then analyzed by six professional Q-graders according to the SCA protocol. Twenty-four hours prior to cupping, the coffee samples were submitted to a light/medium roast (#55 to #65 Agtron color scale). Once the coffee was ground, fragrance and aroma was evaluated. Filtered water (93 °C) was added to the sample cup (five per sample), let to rest for 4 min, and then the beverage was tasted and evaluated according to the quality attributes established in the protocol [5]. Sample classification was based on global scores and aromatic descriptors established by the protocol. It is noteworthy that, given that the goal of this study to evaluate the performance of NIR in comparison to FTIR, the same set of samples was employed for both techniques.

2.2. ATR-FTIR and NIR Analysis

After roasting and grinding, the samples were analyzed on a Shimadzu IRAffinity-1 FTIR Spectrophotometer (Shimadzu, Japan) with a DLATGS (Deuterated Triglycine Sulfate Doped with L-Alanine) detector, using an ATR (Attenuated Total Reflectance) sampling device. The spectra were recorded in the wavenumber range of 3100–800 cm^{-1} and a total of 224 spectra were obtained (56 samples $\times$ 2 aliquots $\times$ 2 measurements). The NIR measurements were conducted in a Red-Wave-NIRX-SD Spectrophotometer (StellarNet Inc, USA) with 25μm diameter and RFX-3D reflectance base. Samples were transferred to a petri dish and placed over this base. The spectra were recorded within 900 to 2300 nm, 16 nm resolution, and 8 scans. Each roasted and ground coffee sample was analyzed in duplicate, totaling 112 spectra (56 samples $\times$ 2 measurements). The background spectra was based on the RS-50 reflectance disk. Both FTIR and NIR analyses were performed at room temperature (20 $\pm$ 0.5 °C) and all readings were based on roasted and ground ($D < 0.15$ mm) coffee samples.

2.3. Data Processing and Statistical Analysis

The software employed for statistical analyses were MATLAB® software v7.9, 2009 (The MathWorks, Natick, MA, USA) and PLS Toolbox® 6.7.1, 2012 (Eigenvector Technologies, Manson, WA, USA). The ATR-FTIR and NIR spectra were used as chemical descriptors in order to build the PLS models for prediction of the sensory analysis scores. The Kennard–Stone algorithm was used to divide the 224 FTIR spectra from FTIR into calibration (70%) and validation (30%) sets, and the same for the 112 NIR spectra. Orthogonal Signal Correction (OSC) and Mean Centering (MC) were applied for reducing the effect of noise, enhancement sample-to-sample differences, and removal of redundant information. The number of latent variables was defined according to the lowest RMSECV value obtained by Random Subset cross-validation. Model performance was measured by calculating the root mean square errors for both calibration (RMSEC) and validation (RMSEP) errors [3]. Selected models were the ones with the smallest RMSEC and RMSEP values [12].

3. Results

3.1. ATR-FTIR and NIR Analysis

The spectra presented in Figure 1 represent the average FTIR spectra of four classes of samples grouped according to their score of sensory quality: 81–83; 84–86; 87–89; 90+. The two bands at the 2900–2850 cm^{-1} range are attributed to C-H vibrations of the bonds present in lipid and caffeine molecules [16]. The marked 1750 cm^{-1} and 1650 cm^{-1} regions are attributed to carbonyl (C=O) vibration and C=C bonds, attributed to carbohydrates and lipids, respectively [20].

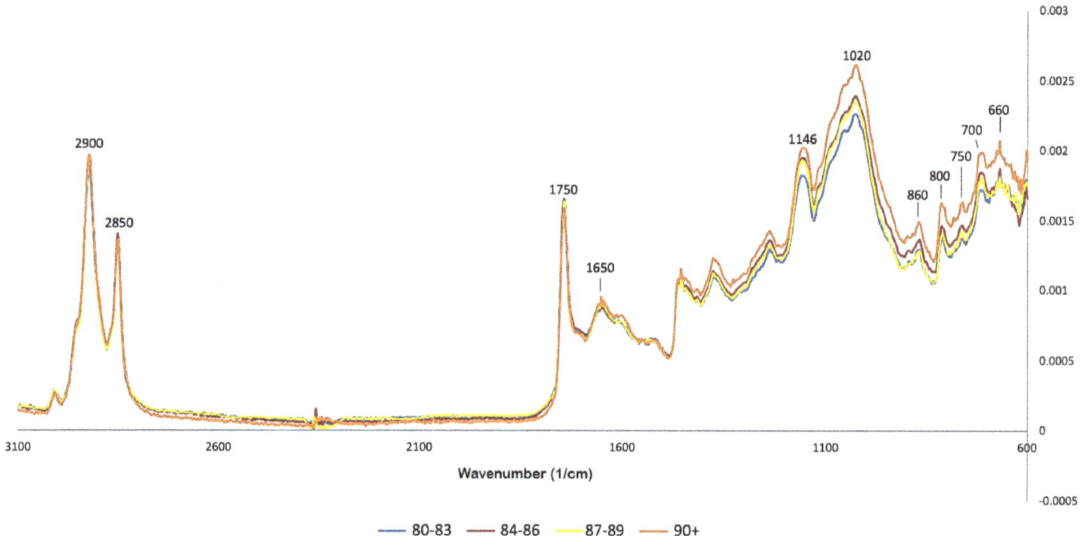

Figure 1. Average FTIR spectra obtained for roasted coffee (colors are related to sensory quality scores).

The bands in the 1650–1600 cm^{-1} range have been previously reported in association with caffeine [4], and were employed in previous studies for quantitative analysis of this substance. The 1680–1630 cm^{-1} range has been found to be associated with vibrations in the carbonyl amide group [4] and also to the presence of trigonelline. The latter is usually decomposed into pyrroles and pyridines during roasting. Pyridines are some of the substances that are responsible for the characteristic aroma of roasted coffee [21].

A significant number of bands can be observed between 1500 and 900 cm^{-1}. Carbohydrates have several absorption bands in the region between 1400 and 900 cm^{-1}, also called "fingerprint region", because it concentrates several relevant bands. The band at 1146 cm^{-1}, has been linked to polysaccharides in previous studies, specifically to the C-O-C stretching of the glycosidic link in the cellulose molecule [4]. Bands in this have also been attributed to amino acids and proteins [22]. Nonetheless, an accurate chemical assignment of bands in this region is still a challenge because of highly coupled vibration modes of polysaccharide backbones [4]. The bands at the 1450–1150 cm^{-1} range have been reported in association with the presence of chlorogenic acids [16,22]. The band at 930 cm^{-1} has been previously reported in association with residues of 3,6-anhydro-galactopyranose [23] resulting from the thermal degradation of polysaccharides, such as galactomannan and arabinogalactan. The levels of chlorogenic acid and trigonelline as well as carbohydrate content will change significantly with roasting, so variations in the fingerprint region of the spectra are expected [14,24].

Figure 2 shows the average NIR spectra of the samples, with different colors being associated with the sensory quality score. The most relevant bands present in the data are as follows: 1100–1250 nm (associated to CH, C-H2, and CH3 overtones from proteins, lipids, caffeine, and organic acids), and 1300–1490 nm (first overtones of RN-H of proteins, first overtones of OH of water and acids) [25]. The band in the 1900 nm region is associated with the combination of O-H angular stretching and deformation, related to the presence of water [26]. The region of 1208–1236 nm is the second bond overtone of C-H, C-H2, and C-H3, as well as the 1700–1720 nm region, which is related to the first overtone of the same carbon and hydrogen bonds, and C-H bonds linked to aromatic rings [26].

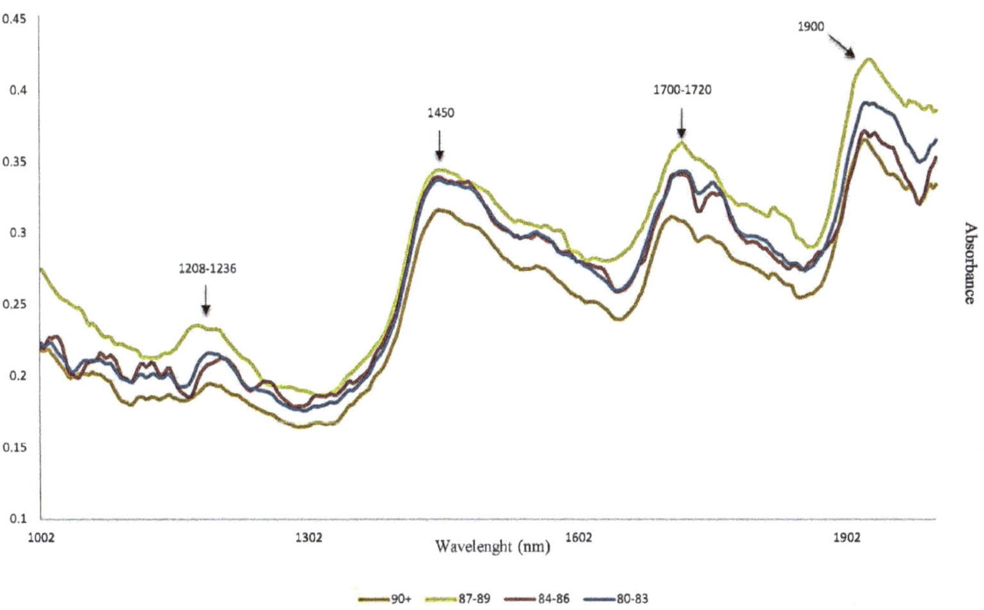

Figure 2. Full NIR spectra (1000–2000 nm) obtained for roasted coffee (colors are related to sensory quality scores.

3.2. Partial Least Squares Regression (PLS)

Figures 3 and 4 show the plots of measured vs. estimated values obtained for the models based on the spectra (estimated) in comparison to the quality scores provided by the Q-graders (measured). The models' parameters are shown in Table 1. The FTIR model was built with two latent variables that were able to explain 99.71% and 81.2% of the accumulated variance in the spectra and sensory data, respectively. Both the values obtained for RMSEP and RMSEC were 0.23%, whereas calibration and validation coefficients were 0.99 and 0.97, respectively. In comparison, the NIR model used three latent variables that explained 90.4% of spectra data variance and 54.05% of the score (sensory) data. The RMSEC value was 0.50% and RMSEP value of 0.52%, and both the calibration (Rc) and validation (Rv) correlation coefficients were 0.98. Although both NIR and FTIR were able to provide good predictions, the FTIR results were slightly more accurate, given the smaller values for RMSEC and RMSEP and slightly higher values for calibration and validation in comparison to NIR. The potential of FTIR as a tool to classify specialty coffees was reported by Belchior et al. [3]. The comparison of both techniques highlights the efficiency of NIR as well. Nonetheless, despite its great potential in food analysis, the interpretation of the spectra in NIR analysis is challenging due to its broadband nature, which consists of overlapping overtone and combination bands [7,9]. The use of NIR-based models represents a new approach in comparing the chemical data with the SCA protocol for coffee classification. Although some studies have reported the use of NIR to evaluate coffees [9,25,27], the discrimination between high quality samples as well as the comparison with the SCA classification shown in this study confirms the potential of this method, providing the coffee industry with a good perspective for using different spectroscopy tools to evaluate coffee quality. Although previous studies [26,27] were able to show that NIR can be used to predict specific coffee sensory parameters (body, acidity, flavor, aftertaste, etc.), this is the first study that extends this application to quantifiable quality scores using an internationally accepted sensory protocol, thus confirming the potential of this method for coffee quality evaluation.

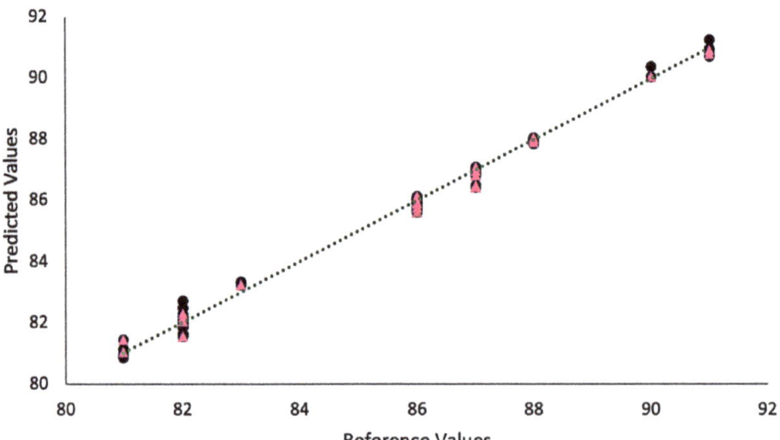

Figure 3. Experimental (black circles) vs. predicted values (pink triangles) obtained by the models based on FTIR spectra.

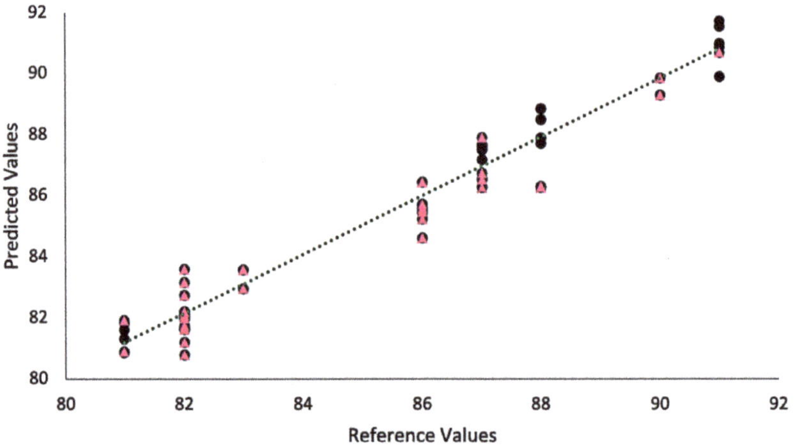

Figure 4. Experimental (black circles) vs. predicted values (pink triangles) obtained by the models based on NIR data.

Table 1. Comparison of the PLS models for both FTIR and NIR techniques.

Model	FTIR	NIRS
Calibration set	149	74
Validation set	67	37
Latent variables	2	3
RMSEC	0.23	0.50
RMSEP	0.23	0.52
Rc	0.99	0.98
Rv	0.97	0.98

RMSEC = root mean square error of calibration; RMSEP = root mean square error of validation; Rc = calibration correlation coefficient; Rv = validation correlation coefficient.

Figures 5 and 6 show the Variable Importance of Projection (VIP) scores of the models. A VIP score is a measure of a variable's importance in the PLS model and is calculated as the weighted sum of squares of the PLS weights, taking into account the amount of explained variance in each extracted latent variable (dimension). Thus, VIP scores above 1 are a typical rule for selecting relevant variables in a given model. An evaluation of the FTIR VIP scores (Figure 5) indicates that the entire spectrum affected the coffee classification in association with the SCA classification. The fact that bands in the whole wavenumber range were relevant in terms of sample classification indicates that many substances that are present in the coffee beverage have significant impact on the sensory profile. Besides coffee being a complex food matrix, several variables that affect the coffee processing chain processes (cultivation, harvesting, post-harvesting, storage, roasting, grinding, and extraction) will impact the final product and affect sensory variations that can be perceived by the Q-graders. Although roasting conditions are consistent and established in terms of the SCA protocol, there still can be variations in the roasting profile (environmental conditions, type of equipment, etc.) [28,29].

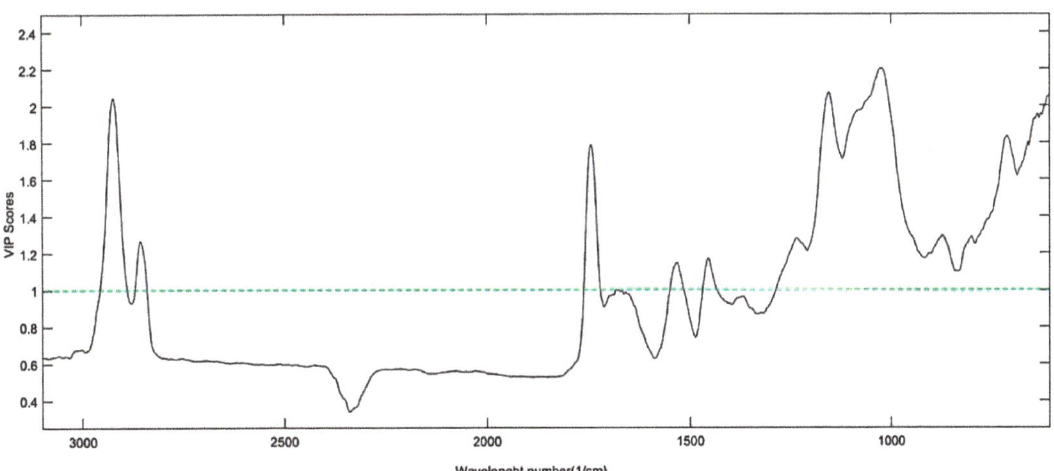

Figure 5. VIP Scores of the PLS models based on FTIR data.

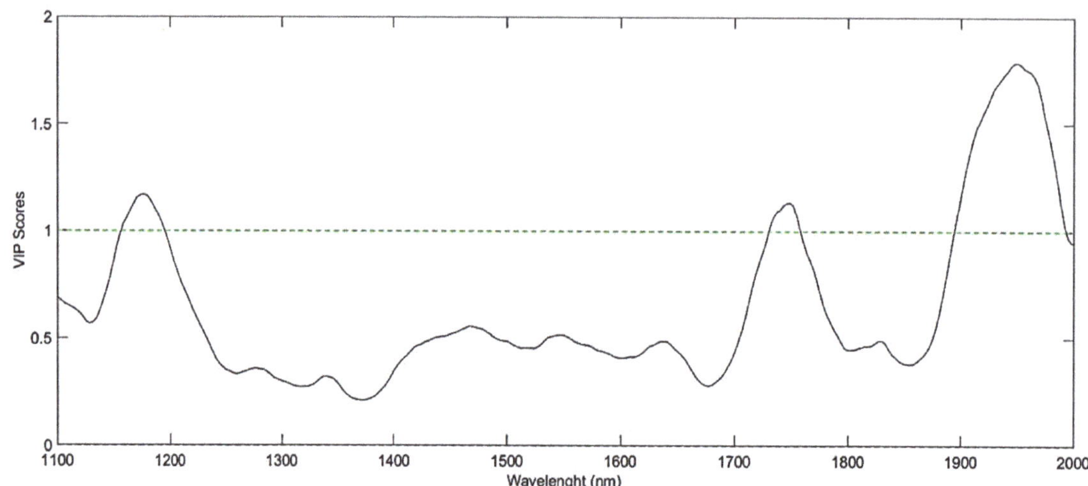

Figure 6. VIP Scores of the PLS models based on NIR data.

VIP scores obtained for the NIR model (Figure 6) show the bands 1176, 1749, and 1950 nm as the most relevant in predicting the coffee scores, being related to the second overtones of CH, C-H2, and CH3, first overtones of CH and C-H2, first overtones of OH, RCO2R, and CONH2, and second overtones of C=O [26]. These regions were assigned by Ribeiro et al. [26] as related to the sensory characteristics of the attributes: taste, cleanliness, and body of the beverage. The region comprised between 1156-1172 nm is attributed to caffeine, and 1738–1755 nm to the presence of lipids in the samples. The region between 1937–1959 nm are related to the ACG and water content of the samples. The beverage cleanliness, regarding the quality of body, is a relevant attribute evaluated in coffee and responsible for higher scores, reinforcing the feasibility of NIR data in adding more confidence to the results.

Schenker and Rothgeb [30] stated that the roasting process can be divided into three stages: drying, Maillard reactions, and development. Therefore, the sensory profile of the roasting coffee will be directly affected by roasting time because it is directly related to the specific phase of the roasting process. This will affect the final coffee composition in terms of several components and reactions, including chlorogenic acids and their derivatives, sugar caramelization, organic acids, volatiles, lipid migration, and melanoidin production; such composition will have a direct effect on the sensory profile [18,31,32]. Therefore, the possibility of validating the sensory analysis performed by Q-graders by using spectroscopic methods is quite relevant. The results obtained in this study are promising for the classification of specialty coffees and confirm the potential of both NIR and FTIR as fast and efficient alternatives for the task at hand. Furthermore, the results are of high interest to the coffee industry, bringing more confidence to the trading routine, given possible inconsistencies between classification of the same samples by sellers and buyers.

4. Conclusions

Spectroscopy-based methods, FTIR and NIR, were shown to be appropriate tools for confirming and predicting score classifications given by Q-graders to roasted specialty coffee samples. The results are promising from the chemometrics standpoint, with models presenting high values for calibration and validation correlation coefficients for both techniques, showing that NIR is also a good tool for predicting coffee quality. It is noteworthy that, even with all samples being of high quality, it was possible to discriminate the nuances in sensory profile. Although the analysis of the whole FTIR spectra of coffee seems to be slightly more efficient from a scientific point of view, NIR spectra also provided robust results related to relevant chemical parameters that define specialty coffee, such as the cleanliness of the beverage. NIR seems promising for routine analysis of specialty coffees, given its simplicity and the possibility of using portable equipment. Therefore, both techniques can be used to confirm and verify the coffee quality scores associated with the Q-graders assessment. As a result, the coffee industry would increase confidence in trading purposes, producing more consistent results. Nonetheless, further studies are needed in order to increase model robustness and applicability, given the intrinsic variations in coffee samples associated to geographical origin, edaphoclimatic conditions, cultivation, and processing techniques as well as variations in roasting parameters. The variability in roasting conditions and equipment in the case of commercially available roasted coffee samples, and the fact that the present methodology was not validated for such conditions, is noteworthy. One of the difficulties in using sensory analysis in the case of the coffee beverage is the need for strict control of roasting conditions in order to guarantee that the tasters will be able to perceive the flavors appropriately. The SCA protocol and the models herein used will be able to correctly classify specialty coffees prior to roasting, but are not suitable for samples that are already acquired as roasted coffees with varying degrees of roast.

Supplementary Materials: The following supporting information can be downloaded at: https://www.mdpi.com/article/10.3390/foods11111655/s1, Table S1: Coffee sample provenance; sensory scores and description provided by the Q-graders.

Author Contributions: Conceptualization, V.B. and A.S.F.; Data curation, V.B. and B.G.B.; Formal analysis, V.B., B.G.B. and A.S.F.; Funding acquisition, A.S.F.; Methodology, V.B., B.G.B. and A.S.F.; Project administration, A.S.F.; Supervision, A.S.F.; Writing—original draft, V.B. and B.G.B.; Writing—review and editing, V.B. and A.S.F. All authors have read and agreed to the published version of the manuscript.

Funding: A.S.F. acknowledges financial support from the Brazilian National Council for Scientific and Technological Development, CNPq (Grant # 310456/2021-5). V.B. acknowledges her scholarship from Coordenação de Aperfeiçoamento de Pessoal de Nível Superior, CAPES.

Institutional Review Board Statement: The study was conducted in accordance with the Declaration of Helsinki, and approved by the Brazilian National Ethics Committee (CAAE code 56961316.0.0000.5093 aproved in 8 August 2016).

Informed Consent Statement: Informed consent was obtained from all subjects involved in the study.

Data Availability Statement: Data is contained within the article (or supplementary material).

Acknowledgments: IKAWA® Sample Roaster Pro (London, UK) supplied by Macchine Per Caffè Ltda (São Paulo, São Paulo, Brazil).

Conflicts of Interest: The authors declare no conflict of interest.

References

1. Cecafé. Available online: https://www.cecafe.com.br/en/publications/monthly-exports-report/ (accessed on 14 March 2022).
2. Yeretzian, C.; Blank, I.; Wyser, Y. Protecting the Flavors—Freshness as a Key to Quality. In *The Craft and Science of Coffee*; Folmer, B., Ed.; Elsevier: London, UK, 2017; pp. 329–353. [CrossRef]
3. Belchior, V.; Botelho, B.G.; Casal, S.; Oliveira, L.S.; Franca, A.S. FTIR and Chemometrics as Effective Tools in Predicting the Quality of Specialty Coffees. *Food Anal. Methods* **2020**, *13*, 275–283. [CrossRef]
4. Craig, A.P.; Botelho, B.G.; Oliveira, L.S.; Franca, A.S. Mid infrared spectroscopy and chemometrics as tools for the classification of roasted coffees by cup quality. *Food Chem.* **2018**, *245*, 1052–1061. [CrossRef] [PubMed]
5. Specialty Coffee Association. Available online: https://sca.coffee/ (accessed on 13 March 2022).
6. Pereira, L.L.; Cardoso, W.S.; Guarçoni, R.C.; da Fonseca, A.F.A.; Moreira, T.R.; Caten, C.S. The consistency in the sensory analysis of coffees using Q-graders. *Eur. Food Res. Technol.* **2017**, *243*, 1545–1554. [CrossRef]
7. Belchior, V.; Franca, A.S.; Oliveira, L.S. Potential of diffuse reflectance infrared Fourier transform spectroscopy and chemometrics for coffee quality evaluation. *Int. J. Food Eng.* **2016**, *2*, 1–8. [CrossRef]
8. Franca, A.S.; Nollet, L.M.L. *Spectroscopic Methods in Food Analysis*, 1st ed.; CRC Press: Boca Raton, FL, USA, 2017. [CrossRef]

9. Munyendo, L.; Njoroge, D.; Hitzmann, B. The Potential of Spectroscopic Techniques in Coffee Analysis—A Review. *Processes* **2021**, *10*, 71. [CrossRef]
10. Wang, N.; Fu, Y.; Lim, L.T. Feasibility Study on Chemometric Discrimination of Roasted Arabica Coffees by Solvent Extractio and Fourier Transform Infrared Spectroscopy. *J. Agric. Food Chem.* **2011**, *59*, 3220–3226. [CrossRef]
11. Reis, N.; Franca, A.S.; Oliveira, L.S. Quantitative evaluation of multiple adulterants in roasted coffee by Diffuse Reflectance Infrared Fourier Transform Spectroscopy (DRIFTS) and chemometrics. *Talanta* **2013**, *115*, 563–568. [CrossRef]
12. Reis, N.; Franca, A.S.; Oliveira, L.S. Concomitant Use of Fourier Transform Infrared Attenuated Total Reflectance Spectroscopy and Chemometrics for Quantification of Multiple Adulterants in Roasted and Ground Coffee. *J. Spectrosc.* **2016**, *2016*, 4974173. [CrossRef]
13. Correia, R.; Cunha, P.; Agnoletti, B.; Pereira, L.; Partelli, F.; Filgueiras, P.; Lacerda, V.; Romão, W. Infravermelho portátil na região do próximo (NIR) aplicado no controle de qualidade de cafés adulterado por borra. *Quim. Nova* **2022**, *45*, 392–402. [CrossRef]
14. Craig, A.P.; Franca, A.S.; Oliveira, L.S. Discrimination between defective and non-defective roasted coffees By Diffuse Reflectance Infrared Fourier Transform Spectroscopy. *LWT* **2012**, *47*, 505–511. [CrossRef]
15. Craig, A.P.; Franca, A.S.; Oliveira, L.S.; Irudayaraj, J.; Ileleji, K. Application of elastic net and infrared spectroscopy in the discrimination between defective and non-defective roasted coffees. *Talanta* **2014**, *128*, 393–400. [CrossRef]
16. Craig, A.P.; Franca, A.S.; Oliveira, L.S.; Irudayaraj, J.; Ileleji, K. Fourier transform infrared spectroscopy and near infrared spectroscopy for the quantification of defects in roasted coffees. *Talanta* **2015**, *134*, 379–386. [CrossRef] [PubMed]
17. Belchior, V.; Botelho, B.G.; Oliveira, L.S.; Franca, A.S. Attenuated Total Reflectance Fourier Transform Spectroscopy (ATR-FTIR) and chemometrics for discrimination of espresso coffees with different sensory characteristics. *Food Chem.* **2019**, *273*, 178–185. [CrossRef] [PubMed]
18. Debona, D.G.; da Silva Oliveira, E.C.; ten Caten, C.S.; Guarçoni, R.C.; Moreira, T.R.; Moreli, A.P.; Pereira, L.L. Sensory analysis and mid-infrared spectroscopy for discriminating roasted specialty coffees. *Coffee Sci.* **2021**, *16*, 1–9. [CrossRef]
19. Hernández-Hierro, J.M.; Cozzolino, D.; Feng, C.H.; Rato, A.E.; Nogales-Bueno, J. Editorial: Recent Advances of Near Infrared Applications in Fruits and Byproducts. *Front. Plant Sci.* **2022**, *13*, 858040. [CrossRef] [PubMed]
20. Abreu, M.B.; Marcheafave, G.G.; Bruns, R.E.; Scarminio, I.S.; Zeraik, M.L. Spectroscopic and Chromatographic Fingerprints for Discrimination of Specialty and Traditional Coffees by Integrated Chemometric Methods. *Food Anal. Methods* **2020**, *13*, 2204–2212. [CrossRef]
21. Illy, A.; Viani, R. *Espresso Coffee: The Chemistry of Quality*, 2nd ed.; Academic Press: Cambridge, MA, USA, 1998.
22. Fioresi, D.B.; Pereira, L.L.; da Silva Oliveira, E.C.; Moreira, T.R.; Ramos, A.C. Spectroscopic and Chromatographic Fingerprints for Discrimination of Specialty and Traditional Coffees by Integrated Chemometric Methods. *Food Control* **2021**, *121*, 107625. [CrossRef]
23. Gómez-Ordóñez, E.; Rupérez, P. FTIR-ATR spectroscopy as a tool for polysaccharide identification in edible brown and red seaweeds. *Food Hydrocoll.* **2011**, *25*, 1514–1520. [CrossRef]
24. Franca, A.S.; Mendonça, J.C.F.; Oliveira, S.D. Composition of green and roasted coffees of different cup qualities. *LWT* **2005**, *38*, 709–715. [CrossRef]
25. Manuel, M.N.B.; da Silva, A.C.; Lopes, G.S.; Ribeiro, L.P.D. One-class classification of special agroforestry Brazilian coffee using NIR spectrometry and chemometric tools. *Food Chem.* **2022**, *366*, 130480. [CrossRef]
26. Ribeiro, J.S.; Ferreira, M.M.C.; Salva, T.J.G. Chemometric models for the quantitative descriptive sensory analysis of Arabica coffee beverages using near infrared spectroscopy. *Talanta* **2022**, *83*, 1352–1358. [CrossRef] [PubMed]
27. Tolessa, K.; Rademaker, M.; Baets, B.D.; Boeckx, P. Prediction of specialty coffee cup quality based on near infrared spectra of green coffee beans. *Talanta* **2016**, *150*, 367–374. [CrossRef] [PubMed]
28. Schenker, S.T.R. Espresso coffee: The chemistry of quality. In *The Craft and Science of Coffee*; Folmer, B., Ed.; Academic Press: Cambridge, MA, USA, 2017; pp. 245–271.
29. Wei, F.; Tanokura, M. Chemical Changes in the Components of Coffee Beans during Roasting. In *Coffee in Health and Disease Prevention*; Preedy, V.R., Ed.; Elsevier: London, UK, 2015; pp. 83–91. [CrossRef]
30. Schenker, S.; Rothgeb, T. The Roast—Creating the Beans Signature. In *The Craft and Science of Coffee*; Folmer, B., Ed.; Elsevier: London, UK, 2017; pp. 245–271.
31. Munchow, M.; Alstrup, J.; Steen, I.; Giacalone, D. Roasting Conditions and Coffee Flavor: A Multi-Study Empirical Investigation. *Beverages* **2020**, *6*, 29. [CrossRef]
32. Febvay, L.; Hamon, E.; Recht, R.; Andres, N.; Vincent, M.; Aoudé-Werner, D.; This, H. Identification of markers of thermal processing "roasting" in aqueous extracts of Coffea arabica L. seeds through NMR fingerprinting and chemometrics. *Magn. Reson. Chem.* **2019**, *57*, 589–602. [CrossRef]

Article

The Elemental Fingerprints of Different Types of Whisky as Determined by ICP-OES and ICP-MS Techniques in Relation to Their Type, Age, and Origin

Magdalena Gajek [1,*], Aleksandra Pawlaczyk [1], Krzysztof Jóźwik [2] and Małgorzata Iwona Szynkowska-Jóźwik [1]

[1] Faculty of Chemistry, Institute of General and Ecological Chemistry, Lodz University of Technology, Zeromskiego 116, 90-543 Lodz, Poland; aleksandra.pawlaczyk@p.lodz.pl (A.P.); malgorzata.szynkowska@p.lodz.pl (M.I.S.-J.)

[2] Faculty of Mechanical Engineering, Institute of Turbomachinery, Lodz University of Technology, Wolczanska 217/221, 93-005 Lodz, Poland; krzysztof.jozwik@p.lodz.pl

* Correspondence: magdalena.gajek@edu.p.lodz.pl; Tel.: +48-42-631-30-95

Abstract: A total of 170 samples of whisky from 11 countries were analysed in terms of their elemental profiles. The levels of 31 elements were determined by Inductively Coupled Plasma Mass Spectrometry (ICP-MS): Ag, Al, B, Ba, Be, Bi, Cd, Co, Cr, Cu, Li, Mn, Mo, Ni, Pb, Sb, Sn, Sr, Te, Tl, U, and V, Inductively Coupled Plasma Optical Emission Spectrometry (ICP-OES) Ca, Fe, K, Mg, P, S, Ti, and Zn and Cold Vapor-Atomic Absorption (CV-AAS): Hg techniques in those alcoholic samples. A comparative analysis of elemental profiles was made on the basis of the content of chosen elements with regard to selected parameters: country of origin, type of whisky (single malt and blended) and age of products. One of the elements which clearly distinguishes single malt and blended types of whisky is copper. Single malt Scotch whisky had a uniform concentration of copper, which is significantly higher for all malt whisky samples when compared with the blended type. Analysis of samples from the USA (n = 26) and Ireland (n = 15) clearly revealed that the objects represented by the same product but originating from independent bottles (e.g., JB, JDG, BUS brands) show common elemental profiles. On the other hand, comparative analysis of Scotch whisky with respect to aging time revealed that the longer the alcohol was aged, (i.e., the longer it stayed in the barrel), the higher the content of Cu and Mn that was recorded.

Keywords: whisky; elemental analysis; ICP-MS; ICP-OES; CV-AAS; spirits; PCA; metals

Citation: Gajek, M.; Pawlaczyk, A.; Jóźwik, K.; Szynkowska-Jóźwik, M.I. The Elemental Fingerprints of Different Types of Whisky as Determined by ICP-OES and ICP-MS Techniques in Relation to Their Type, Age, and Origin. *Foods* **2022**, *11*, 1616. https://doi.org/10.3390/foods11111616

Academic Editor: Daniel Cozzolino

Received: 5 April 2022
Accepted: 26 May 2022
Published: 30 May 2022

1. Introduction

Whisky (whiskey—alternate spelling is commonly used in Ireland and the USA—for consistency, the former spelling is used in this paper) is one of the most popular high-percentage alcoholic beverages made from grain in the world. In accordance with the present definition, whisky is a kind of distilled spirit made from fermented grain mash. Many types of whisky are associated with various types of production. In the case of European products, the alcohol should be matured for at least 3 years in wooden barrels of a volume not exceeding 700 L, and only water and caramel (for colouring) can be added to the distillate. For example, similar requirements for the type of grain from which whisky is produced are applied to both Scottish and Irish beverages. However, in Scotland, double distillation is used, while in Ireland it is tripled. The possibility for adding exogenous amylolytic enzymes in the mashing process for Irish whisky is another difference. In turn, alcohol produced in the USA (American Bourbon) is most typically aged less than 4 years (e.g., 2 years for the European market). Furthermore, bourbon in the USA has to be produced from a mixture of grains consisting of no less than 51% corn. The distillate must contain no more than 80% pure alcohol. Moreover, maturation takes place in new oak barrels, fired from the inside, which significantly differentiates this process from the one used in the production of European whisky, which is matured in previously used barrels (after wine, bourbon, or beer maturation process). In the USA, the top producers of bourbon have their distilleries in Tennessee and Kentucky. Alcohol branded as "Tennessee whisky" is known to have been subjected to a 10-day purification process using a layer of charcoal prepared from maple wood [1–3].

The analysis of whisky, both in terms of chemical composition characterisation and authentication, mostly involves the employment of separation techniques, such as gas or liquid chromatography, often coupled with flame ionization detection (FID) or mass spectrometry (MS). It needs to be highlighted that these techniques are mostly applied as a target type of analysis, where specific markers are traced and determined within the whisky authenticity verification [4–8]. Apart from VOCs and other organic compounds that, determined by the chromatographic methods, can be used as indicators for the identification of origin, alcoholic beverages can be also tested for trace elements, which are derived from the raw materials, production process equipment, storage vessels, and additives. Compared with those on organic markers, studies on trace elements used for the identification of counterfeit whisky are very limited. The first attempts to determine metals in whisky samples were made in 1998. Anodic stripping voltammetry (ASV) and atomic absorption spectroscopy (AAS) were applied at that time to measure and compare the levels of Zn, Pb, and Cu in four whisky samples. The authors noted that the stripping method had an important advantage over AAS in terms of lower detection limits. In the case of ASV, these limits were 4, 18, and 100 times lower for Zn, Cu, and Pb respectively. This is a key problem for heavy metals (Pb) since it is impossible to measure them using the AAS technique [9]. In 2002, Adam et al. [10] conducted trace elemental analysis on 35 Scotch whisky samples to verify whether there were trace element fingerprints characteristic of different kinds of Scotch whisky. A total of 31 samples of single malts, 1 sample of malt blend, 2 samples of blended Scotch, and 1 sample of grain whisky were analysed. For only the measurement of copper, an additional number of whisky samples was studied (6 blended Scotch whiskies, 11 single malt whiskies, and 1 rye whisky). The samples were taken directly from whisky bottles purchased from a supermarket. The selected malts originated from 4 Scottish regions: the Lowlands, the Highlands, Speyside, and Islay, and were aged between 6 and 20 years. For the determination of the selected metals, a graphite furnace atomic absorption spectrometer (GFAAS) was employed. The authors of the paper stated that the fingerprint of the metal concentrations in whisky could not be used as a criterion to identify whiskies from different production regions. However, when the second set of samples (42 malt whiskies and 8 blended whiskies) was analysed for copper, the concentration of this element could have been, according to authors, used as a criterion to distinguish blended or grain Scotch whiskies from malt whiskies. Much higher levels were observed in the malt whiskies in comparison with the concentration of copper in the blended Scotch whiskies or the pure grain whiskies. The authors concluded that the difference between these levels was highly significant. It was suggested that the copper analysis itself could be used as one of the markers to distinguish between blended and single malt Scotch whiskies. The main sources for the presence of copper in whisky are the copper stills in which whisky is distilled and the barrels in which the spirits are aged. Additionally, the authors indicated a possible relationship between the copper content and the acidity of the alcohol. In 2017, Shand et al. [11] made an attempt to use the elemental analysis of Scotch whisky performed by total reflection X-ray fluorescence as a potential tool in the identification of counterfeits. Elements such as Cu, Zn, Fe, Ca, S, Cl, K, Mn, P, Rb, and Br were selected because their presence is associated with the whisky production process. Moreover, their concentrations in most samples were above the limits of detection offered by TXRF. In total, 32 samples were analysed, of which 17 were single malt whiskies produced in different regions of Scotland (the Highlands, the Lowlands, Speyside, and Islay), 8 samples were blended scotch whiskies, and 2 were grain whiskies. Additionally, 5 samples were counterfeit whiskies from sources which remained anonymous. The samples were analysed without any special preparation process. A total of 18 out of the 32 were also checked by ICP-OES (after earlier sample preparation by the evaporation to dryness and the addition of nitric acid). In order to discriminate between the whisky samples in accordance with the indicated parameters, the authors used multivariate analysis. The principal component analysis (PCA) indicated that the counterfeit samples could be distinguished from the others on the basis of their trace elemental profiles. The second component was especially important in separating the counterfeit samples from the authentic Scotch whiskies. In turn, the third component had the greatest impact on the separation of classes (Highland, Lowland, Speyside, Islay, blended, grain, and counterfeit). The authors also observed statistically significant and strong positive correlations between Rb, K, and Mn. However, there was no obvious chemical or geochemical connection between these elements, which was underlined by the authors. Additionally, the applied CA analysis showed the unambiguous grouping of counterfeit samples. The linear discriminant analysis (LDA) made it possible in most cases to correctly classify the studied whisky samples into appropriate groups. It was extremely important, especially for the counterfeit samples.

Due to the fact that only a few, limited papers on metal analysis of whisky are available, the main goal of the authors was to perform an extensive, elemental characterization of whisky samples. Moreover, the possibility of using statistical analysis and chemometric tests to differentiate and distinguish whisky samples, based on their origins, types, and ages, was tested. Therefore, the present work presents an extremely rare approach to the assessment of selected whisky parameters, based on extensive elemental analysis. It should be emphasized that, in this study, a wide range of measurements was carried out with the use of 3 analytical techniques (ICP-MS, ICP-OES, and CV-AAS) to determine the concentrations of 31 elements in 170 whisky samples.

2. Materials and Methods

2.1. Samples

In total, 170 whisky samples (152 various brands of single malt and blended, high-percentage alcoholic beverages) originating from 11 countries (Scotland, the USA, Ireland, Poland, Japan, the United Kingdom, India, Azerbaijan, Slovakia, Wales, and Bulgaria) were chosen for elemental analysis using the by Inductively Coupled Plasma Mass Spectrometry (ICP-MS), Inductively Coupled Plasma Optical Emission Spectrometry (ICP-OES), and Cold Vapor-Atomic Absorption (CV-AAS) techniques. Alcohol samples selected in this study consisted partially of brands of whisky widely available on the Polish market, which can be found in the supermarkets. Some of the samples were obtained through official whisky distributors. Some of them are distillates intended for concentration and are thus not available for direct sale. The names of the whiskies are coded, and the manufacturers' names are not given in this paper. The basic characteristics of the tested samples are included in Table 1.

Table 1. Characteristics of the tested set of samples.

N	Scotland	USA	Ireland	Poland	Others	
	Single Malt—50 Blended—56	Single Barrel—1 Blended—25	Single Malt—3 Blended—12	Single Malt—3 Blended—7	Japan—3 India—3 Slovakia—3	UK—1 Azerbaijan—1 Wales—1
					Bulgaria—1	
Total	106	26	15	10	13	

2.2. Samples Preparation and Equipment

- ICP-OES and ICP-MS

The following measurement techniques were applied in this study: ICP-OES (Thermo Scientific, ICAP 7000 series, Bremen, Germany) and ICP-MS (Thermo Electron Corporation, X SERIES, East Lyme, CT, USA). These techniques required that the samples be prepared in a decomposed form, which process was performed in a microwave system (Ultrawave system, Milestone, Via Fatebenefratelli, Italy). For this purpose, 4 mL of 69–70% HNO_3 (Baker, Avantor Performance Materials Poland S.A., Gliwice, Poland) were added to 4 mL of each of the samples. The acid was added in small portions due to the strongly exothermic nature of the reaction. In the next step, microwave mineralization was employed. The procedure was analogous to that used on the whisky samples described in our 2019 preliminary study [12]. After the mineralization process, the contents of the tubes were quantitatively transferred to flasks with a volume of 25 mL. A standard of In with a certified concentration was used as an internal standard to monitor signal stability (Merck, Warszawa, Poland). For the measurement of the indicated elements, it was necessary to prepare calibration curves based on a standard solution of CPAchem (Multi-element ICP standard, Stara Zagora, Bulgaria), and some single-element standards of In (ICP class, Merck, Darmstadt, Germany), Sb (ICP class, Merck, Darmstadt, Germany), Sn (ICP class, Chem Lab NV, Zedelgem, Belgium), Ti (ICP class, Radian International LLC, Austin, TX, USA), S (ICP class, Merck, Darmstadt, Germany), and P (ICP class, SCP Science, Québec, Canada). The preparation of the standards was carried out by the subsequent dilution method. The blank samples were prepared in the same way as the studied samples.

An ICP-MS analytical technique was applied to determine the levels of metals in the whisky samples based on the following isotopes: ^{107}Ag, ^{27}Al, ^{11}B, ^{138}Ba, ^{9}Be, ^{209}Bi, ^{111}Cd, ^{59}Co, ^{52}Cr, ^{63}Cu, ^{7}Li, ^{55}Mn, ^{95}Mo, ^{60}Ni, ^{208}Pb, ^{121}Sb, ^{118}Sn, ^{88}Sr, ^{125}Te, ^{203}Tl, ^{238}U, and ^{51}V. In turn, concentrations of Ca (393.366 nm), Fe (238.204 nm), K (766.490 nm), Mg (279.553 nm), P (185.942 nm), S (180.731 nm), Ti (334.941 nm), and Zn (213.856 nm) were determined by the ICP-OES technique. Information about the operating conditions for the elemental analysis of the whisky samples performed using the ICP-OES and ICP-MS spectrometers is presented in Table S1 in the Supplementary Materials.

A total of three replicates were made for each alcohol beverage sample and analytical technique. The RSD, expressed as a percentage, even for elements measured at very low levels, was in the range of 0.01–5.00%. The accuracy of the applied procedure was verified based on the analysis of the certified reference material of TMDA 54.6 (a fortified lake-water sample from the National Water Research Institute, Burlington, Halton, ON, Canada). The obtained recoveries were close to 100%. The same procedure for verifying the accuracy of the proposed method was described previously by Gajek et al. in 2021 [13].

The coefficient of linear regression for each analyte was in the range from 0.999 to 1.000. The sensitivity of the developed method was considered in terms of the limit of detection (LOD) and the limit of quantification (LOQ). The two limits were based on values of the standard deviation of the results obtained for a series of blank samples, according to the following mathematical expressions: $LOD = x_{śr} \cdot 3SD$ and $LOQ = 3 \cdot LOD$ [14]. The obtained results are presented in Table S2 in the Supplementary Materials.

- CV-AAS

In this study, an automatic mercury analyser MA-3000 (Nippon Instruments Corporation, Tokyo, Japan) was applied to determine the total mercury content in the whisky samples. The analytical procedure was analogous to the one described in detail in 2019 [12].

2.3. Data Analysis

Statistica 12.5 (New York, NY, USA) software was used for the statistical and multivariate analysis. In order to verify the normality of the distribution of the studied variables, Kołmogorow–Smirnow tests were used. Kołmogorow–Smirnow tests are very helpful in the verification process if a sample originates from a population with a specific distribution based on the distance between the empirical distribution function of the sample and the cumulative distribution function of the reference distribution. The application of these tests for the significance level $\alpha = 0.05$ showed that the hypothesis of a normal distribution for all analysed variables (the concentrations of 30 elements) should be rejected. For this reason, the Kruskal–Wallis non-parametric test was used to assess the significance of differences in the determined levels of elements among particular groups according to the parameters considered, such as country of origin, type, and year. The test determines whether the medians of two or more groups are different. The quantitative data were expressed in this study in the form of the box and whisker plots with a median value chosen as a central value. A total of 50% of the most common results are within the box, while the whiskers are limited by the highest and lowest results obtained in this work. To increase the interpretability of the results, multivariate analysis, namely principal component analysis (PCA), was applied. PCA is the basis of multivariate data analysis based on projection methods. The most important application of PCA is to represent multivariate data as a smaller set of variables in order to observe trends, clusters, and outliers. This analysis may uncover the relationships between observations and variables, and among variables themselves.

3. Results and Discussion

3.1. Levels of Metals in Analysed Whisky Samples

In this study, the levels of 31 elements in 170 whisky samples were determined. The concentrations of Ag, Al, B, Ba, Be, Bi, Cd, Co, Cr, Cu, Li, Mn, Mo, Ni, Pb, Sb, Sn, Sr, Te, Tl, U, and V were measured by the ICP-MS technique, but to assess the level of Ca, Fe, K, Mg, P, S, Ti, and Zn, the ICP-OES technique was used. The Hg content was analysed by the CV-AAS technique. In the collected data set, some of the obtained results were below the quantification limits. Te was not quantified in the majority of samples—157. Ag was not detected in 102 samples, nor Sb in 90, nor Ti in 61. Fe was not detected in 53 samples, Zn in 48, P in 34, V in 33, nor Mo in 31 samples. Cd was not determined in 20 samples, Sn in 11, Bi in 10, Tl in 9, nor K in 8 samples. U and Al were not detected in 5 independent samples, B in 2, and neither Pb nor Be was quantified in 1 sample. In the case of mercury content, all results were below the limit of quantification. Thus, this element was excluded from further calculations.

The first step in the work of data processing was to check the hypothesis about the type of distribution. For this purpose, the Kołmogorow–Smirnow test was used to verify the distribution of all analysed samples at the accepted level of significance, $p = 0.05$. The null hypothesis regarding the normal distribution for all variables was rejected. Therefore, the nonparametric Kruskal–Wallis test was used to further analyse the data. The basic statistical information on the studied variables such as the mean, median, minimum, and maximum, has been placed in Table 2.

Table 2. Basic statistics for determined elements for all whisky samples (n = 170) [μg/L].

Element	n	Mean	Median	Min	Max	Element	n	Mean	Median	Min	Max
Ag		4.270	<LOQ	<LOQ	399.1	Sb		3.470	<LOQ	<LOQ	227.9
Al		117.7	113.3	<LOQ	399.7	Sn		9.800	4.670	<LOQ	44.50
B		4388	4116	<LOQ	12.89	Sr		47.18	45.81	15.84	119.2
Ba		188.7	182.4	38.68	950.9	Te		0.040	<LOQ	<LOQ	1.200
Be		0.100	0.090	<LOQ	0.300	Tl		0.110	0.040	<LOQ	2.600
Bi		1.310	0.870	<LOQ	19.80	U		0.260	0.230	<LOQ	0.900
Cd		1.260	0.720	<LOQ	16.00	V		2.210	0.960	<LOQ	57.30
Co	170	4.530	2.470	0.406	74.90	Ca	170	14.66	9185	723.8	175.35
Cr		153.4	111.1	10.70	666.1	Fe		166.6	88.03	<LOQ	1485
Cu		473.7	216.0	16.25	5252	K		18.50	12.45	<LOQ	149.30
Li		21.36	12.27	0.474	399.5	Mg		1487	1046	208.5	11,548
Mn		47.43	32.95	4.396	286.5	P		1637	313.7	<LOQ	30.11
Mo		1.790	1.070	<LOQ	32.30	S		7126	4648	296.3	69.91
N		24.01	12.96	3.201	301.3	Ti		25.68	12.72	<LOQ	288.3
Pb		15.82	10.61	<LOQ	450.9	Zn		1221	177.5	<LOQ	31,458

Despite the quality control of food products prior to their introduction into the market, both reports in the literature and earlier research conducted by authors of this paper [13,15] clearly indicate that the permissible standards can be exceeded. Based on the results obtained for low-percentage alcoholic beverages, such as wines or ciders, cases where both the international and national standards have been exceeded can be found in the literature.

The results obtained for 180 samples of wine studied by Gajek et al., 2021 [15], revealed that, in the case of 18 wine samples, the maximum levels of some metals (Cd—8 samples, Pb—9 samples, and Cu—1 sample) were slightly exceeded according to the OIV standards [16]. A similar observation was found in a study by Woldemariam et al., 2011 [17], where, especially in the case of lead, significant exceedances in wines (from the Czech Republic—max content 1253 μg/L) were reported. On the other hand, in the case of the analysis of cider samples [13], the authors emphasized that, for elements such as Cd and Pb, the maximum obtained results exceeded only the standards for drinking water [18]. The standards for alcoholic beverages were maintained for both elements.

In terms of the elemental whisky analysis, none of the authors dealing with this topic verified the potentially negative impact of exceeding the permissible maximum levels of the selected metals. The authors of this paper referred only to the internal national standards that define the maximum permissible content of the selected metals (Cd and Pb) in high-percentage alcohols [19]. The maximum lead content was set at 0.3 mg/L, and the maximum cadmium content at 0.03 mg/L. In this study, the permissible Pb level was exceeded only for 1 out of 170 analysed whisky samples (the blended whisky from Ireland—max content 450.9 μg/L). The limit value for Cd was not exceeded in any case.

3.2. Elemental Analysis for Country of Origin

3.2.1. General Characteristics

So far, numerous attempts have been made to correlate the chemical composition of whisky with its origin. Most scientists have used chromatographic (GC and HPLC) and spectrophotometric (UV–Vis) techniques along with complex mathematical models to assess the correlation between an alcohol's composition and its geographical origin. In some cases, authors have stated that the conducted measurements did not provide sufficient information to distinguish among whisky samples with regard to their countries of origin [20]. On the other hand, other literature reports clearly suggest that it is possible to distinguish Irish whisky from Scotch and bourbon on the basis of a few markers determined by chromatographic techniques [21]. Other researchers, having only 11 samples of whisky of various origin, were able to discriminate amongst all 5 of the alcohol groups under consideration using the Headspace mass spectrometry technique [22]. So far, there are single scientific studies in which researchers have made an attempt to distinguish whisky origin using multi-elemental analysis. Adam et al., 2002 [10], stated that the fingerprint of the metal concentration of whisky cannot be used as a criterion for identifying whiskies from different Scottish production regions. The authors of this paper also presented similar considerations in their preliminary research [12]. In the 20 tested samples of whisky, originating from different countries (Scotland, the USA, and Ireland), no statistically significant differences were found in any of the cases although the authors indicated that some isotopes (^{48}Ti, ^{138}Ba, ^{66}Zn, ^{90}Zr, and ^{118}Sn) created the characteristic "fingerprint" of the Irish-made whisky sample. On the other hand, the copper content based on isotope ^{63}Cu was considered as crucial in distinguishing the type of whisky (single malt and blended).Additionally, it turned out to be impossible, based on the collected outcomes, to distinguish between various production regions of Scotland (the Lowlands, the Highlands, Speyside, and Islay). However, as the authors emphasize, the number of analysed samples could be too small for the aforementioned comparison to be carried out correctly.

On the basis of the Kruskal–Wallis test, the existence of statistically significant differences in the concentration of the following elements was demonstrated: Ag, Be, Bi, Ca, Cd, Cu, Fe, Li, Mn, P, Sb, Sn, Ti, V, and Zn (Table 3). In the case of all the mentioned elements, the existence of statistically significant differences was confirmed based on the level of significance (p), which was less than 0.05. The most important statistical information connected with the division of samples by country of origin is included in Table S3 and Figure S1A–P (Supplementary Materials). What should be highlighted is the fact that, for most of the elements for which the existence of statistically significant differences were confirmed, whisky originating from Scotland was listed in the majority of comparisons. Although whisky samples from Scotland were represented by the largest number of samples (106), only in the case of Cu and Cd were the highest median values were observed for these elements among all other studied groups. In turn, with the exception of Cu, Cd, Sn, Bi, and Ca, for the rest of the mentioned elements (Li, Be, V, Mn, Ag, Sb, Zn, P, Fe, and Ti) the whisky from the USA was characterized by the highest median values when compared with samples from other countries. However, in most comparisons, copper was one of the crucial elements presented.

Table 3. Groups for which statistically significant differences were reported.

Statistically Significant Differences	Elements
SCT–USA	Li; Be; V; Cu; Ag; Sn; Sb; Zn; P
SCT–IRL	Mn; Cu; Cd
SCT–PL	Sn
SCT–OTH	Bi; Cu
USA–PL	Ca
USA–OTH	Bi; Fe; Ti; Cu
IRL–OTH	Cu

SCT—Scotland; USA—United States of America; IRL—Ireland; PL—Poland; OTH—other countries.

Considering the order of concentrations of the studied elements for which statistically significant differences were confirmed, only for the selected metals the same tendencies can be observed; for example, for Fe and Ti, the following order for median values can be noted: OTH > PL > SCT > IRL > USA. For Zn and V, on the other hand, the order was as follows: SCT > OTH > IRL > PL > USA. In general, for elements such as V, Sn, Zn, Sb, and P, the lowest values were determined in samples from Scotland, while the highest ones were found for products from the USA. Not surprisingly, the lowest content of Cu was characteristic for the whisky from the USA, whereas the alcohol from Scotland had the highest level of this element. For the rest of the studied elements, no common order in relation to the country of production can be indicated. The observed differences only prove that samples from various countries have completely different elemental fingerprints.

3.2.2. Characteristics of Samples from the USA

In the next steps, the samples originating from different countries will be discussed separately (USA—United States of America; IRL—Ireland; PL—Poland; OTH—other countries). The research objects from the USA consisted of 26 samples (each of the samples was coded.) Almost all of the samples from the USA were blended products only one of the tested samples was the single-barrel type of bourbon—BlaSB). Within this group, 12 independent brands were distinguished. The most numerous were the samples of the JB brand, which included 6 products (where JB1, JB2, and JB3 were samples of the same product, coming from different bottles, purchased in different stores during some period of time). The JD brand, which included 5 products, was also represented by a quite large group (where JDG1 and JDG2 were samples of the same product, coming from different bottles, purchased in different stores during some period of time). Moreover, only one of the studied group of samples was a product with an age declaration (JBB6YO). The remaining products were aged for the minimum period of time required by law. The projection of cases on the factor-plane which was made for this group clearly revealed one outlier point (JDS, from the JD brand). This was a limited-edition sample of a well-known brand of whisky from the USA. It should be emphasized that this sample was characterized by the highest content of the following elements in relation to the group under consideration: Li, Co, Ni, Cd, Sn, Tl, Bi, Zn, and P. In order to improve the readability of the graph and obtain a more accurate scale, this point was omitted from Figure 1.

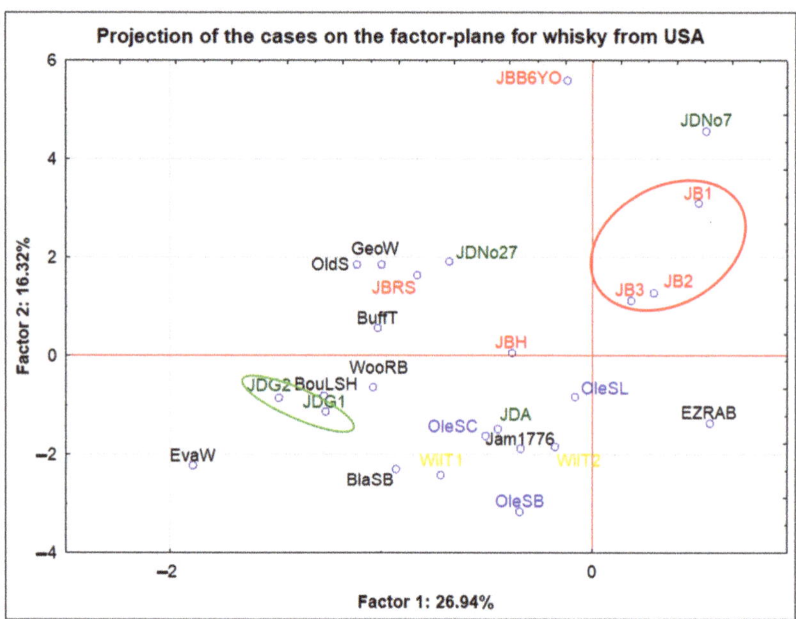

Figure 1. Projection of the cases on the factor-plane for 25 samples (after scale change) from the USA.

The JDG1 and JDG2 (samples from independent bottles of JDG—marked in green) objects from the same brand, JD, in the projection shown in Figure 1, were placed next to each other. However, the remaining JD brand objects (namely JDA, JDNo7, and JDNo27) were dispersed, and therefore, no common elemental features were observed for them.

Objects JB1, JB2, and JB3, being the same product but coming from independent bottles (JB brand—marked in red), also very clearly created a common group in the presented projection. Objects JB2 and JB3 were very close to each other, while object JB1 was slightly shifted. This was probably due to the fact that the JB1 sample came from a bottle from a completely different production batch (the oldest one from the point of view of the purchase date). In relation to the group of objects JB1-3, the other JB brand objects were outlier points. They differed in their aging period in the barrel (JB6YO) or in the addition of flavouring substances (JBH and JBRS).

Although, in the case of the remaining brands represented by more than one sample (e.g., Ole and WilT), it cannot be concluded that they formed common clusters characteristic of the brands, it should be emphasized that these samples were located in one quadrant (III) of the projection of cases on the factor-plane.

Additionally, the existence of statistically significant differences within the whisky samples from the USA was checked by the non-parametric test. The samples were divided according to the producers. Five groups were distinguished (JB, JD, Ole, WilT, and Oth, with the Oth group containing the rest of the single objects). The existence of statistically significant differences was revealed for Mg, Mn, Mo, Pb, and U. Among the mentioned elements, no statistically significant differences were found for Cu contents since the level of this metal was the lowest in samples originating from the USA. For magnesium, the differences concerned the JD and Ole brands, as well as the JD brand and other samples. Similarly, in the case of manganese, differences were noted for the JD brand and the group of other samples. For molybdenum, the differences concerned the JB and Ole brands, as well as the JB brand and other samples. For lead, a difference was found between the JD and Ole brands, and for uranium between the JB and Ole brands. It should be noted that the JD brand was distinguished by the lowest values of Mg and Mn in relation to the other groups. On the other hand, the JB brand was characterized by the highest Mo and U contents in relation to the others. The most important statistical information connected to the division of samples against the brands produced in the USA is included in the Supplementary Materials (Table S4).

3.2.3. Characteristics of Samples from Ireland

Samples of whisky from Ireland included 15 subjects. Within this group, 6 independent brands were distinguished. The most numerous were the samples of the Bus brand (where Bus1, Bus2, and Bus3 were samples of the same product, coming from different bottles, purchased in different stores during some period of time). The projection of cases on the factor-plane which was made for this group clearly revealed three outlier points. Two of them belonged to the Jam brand (each of these alcohols was aged in a different barrel.) The last outlier is Southern Ireland's blended whisky. The drink is a combination of 4-year-old barley distillates with 3-year-old grain distillates. Again, in order to improve the readability of the graph, these points were omitted from Figure 2.

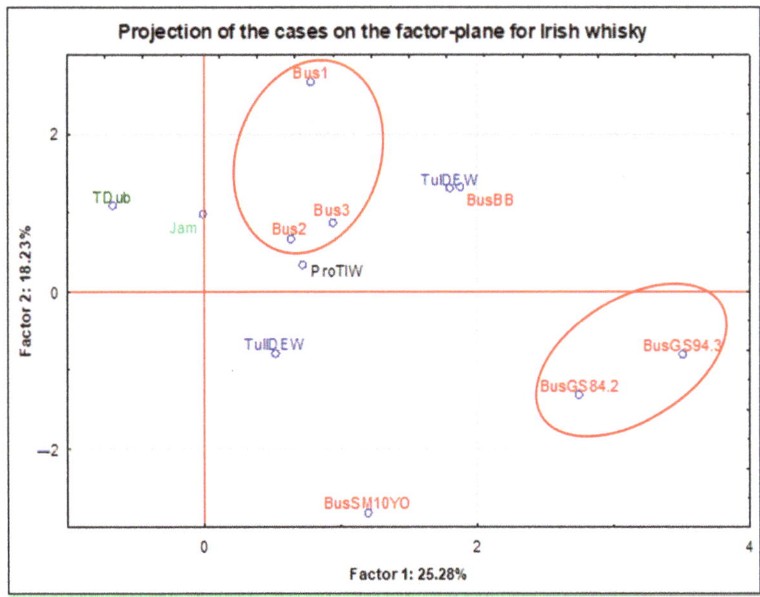

Figure 2. Projection of the cases on the factor-plane for 12 samples (after scale change) from Ireland.

The conducted projection revealed that the objects derived from the same product (Bus1, Bus2, Bus3—marked in red) have a uniform elemental profile, thanks to which, they form a common group. These samples were characterized by the highest content of barium compared to the other samples from Ireland. As with the products from the USA, objects Bus2 and Bus3 were very close to each other, while object Bus1 was slightly shifted. This was probably due to the fact that the JB1 sample came from a bottle from a completely different production batch (and oldest in terms of time of its purchase). BusGS84.2 and BusGS94.3 (marked in red) were samples of single-grain distillates from the Bus brand. These distillates were high-percentage alcohols without an aging process. These were unique samples obtained directly from the distillery, provided by one of the Polish distributors. Again, it can be concluded that the samples of these alcohols had a very consistent elemental profile. They were characterized by a higher content of copper, chromium, and nickel than the others. The other 2 samples from the Bus brand were the 10-year-old single malt (Bus10YO) and the premium-class blend consisting of 75% malt whisky (BusBB). The 10-year-old single malt (Bus10YO) sample seemed to be particularly interesting in terms of elemental composition. It was characterized by the highest values in relation to the other samples from Ireland in terms of the following elements: V, Mn, Ni, Cu, Sr, Sn, and P. As in the case of the JB brand from the USA, the sample with the declared aging period was clearly the outlier. Similarly, in this case, the BusSM10YO object was the only one of the Bus brand products with a declared aging period. This suggests that time may be the most important distinguishing factor (as opposed to brand or origin).

Moreover, as in the case of the product from the USA, the samples from Ireland were divided according to the producers (Bus, Jam, Tul, and Oth: the rest of the single objects). On the basis of the Kruskal–Wallis test, the existence of statistically significant differences in the concentration of B was demonstrated between the group of the products of the Jam and Bus brands. The Jam brand was characterized by a higher content of this element. The most important statistical information for B is included in the Supplementary Materials (Table S5).

3.2.4. Characteristics of Samples from Poland

The set of whisky samples from Poland contained 10 objects, including 8 different brands. Poland is certainly not a country associated with whisky production. However, in recent years, due to rapidly growing consumption, products from domestic brands have been appearing on the market. Most often, producers of vodkas, liqueurs, or mead, wanting to expand their product range, have introduced whisky produced from local ingredients to their portfolios. There are also beverages on the Polish market that are advertised as Polish products but created in cooperation with manufacturers from other countries (most often Scotland). Frequently, they are blended types of whisky made of Scottish barley malts and Polish distillates from other cereals.

The conducted projection of the cases on the factor-plane revealed that the samples from the same manufacturers (WolDS and WilFO) were grouped together (Figure 3). In the case of samples from the WilFO distillery (marked in red), one was a single-grain (WilFOSG), and the second was a single-malt wheat (WilFOSMW). However, the same production method and distillation equipment ensured, in this case, a coherent elemental profile. Moreover, both products from this brand were aged for a period of 3 years. Characteristic features of the WilFO brand were the highest Sn and Pb contents compared to other products from Poland. In turn, samples from the WolDS distillery (marked in navy blue) are single rye whiskies. The main difference between them is the type of barrel in which they were matured. The sample of WolDSRRF was aged in rum barrels, whereas the whisky coded as WolDSRRPOF was matured in a barrel made of Polish oak. As in the case of the previous brand, the maturation period was 3 years. It should be emphasized that the WolDS brand is distinguished from the others due to its having the highest values of Mn and Mg.

An interesting position in the compared group of samples was PolWS (a Polish single malt whisky—marked in dark green). It was produced at home, but according to the definition, it met all the requirements for this type of alcohol. This whisky was aged for 3 years in oak barrels, fired from the inside. What distinguished this sample was its having the highest content of Sr, K, S, and P compared to the other Polish products. Potassium was indicated by Gajek et al., 2021 [13], as an element which much greater content characterizes home-made products.

Among the 10 analysed samples from Poland, 5 were single malt, single rye, or single grain and were located in the projection of the cases on the factor-plane at the top of the plot (quarters I and II). The remaining 5 samples were blended-type products. These points were at the bottom of the projection (quarters III and IV). However, in the case of all 5 blended samples, despite the fact that the manufacturer declared the Polish origin of these products, it was extremely difficult to trace them back to their real origin.

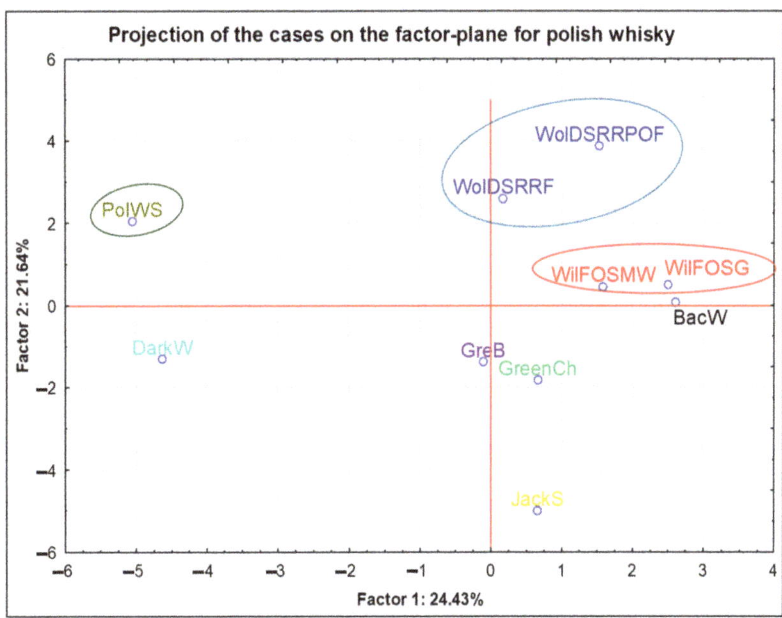

Figure 3. Projection of the cases on the factor-plane for the 10 samples from Poland investigated in this study.

3.2.5. Characteristics of Samples from Scotland

The group of products from Scotland included 106 whisky samples (50 single malt and 56 blended whiskies). The studied objects in this group of products were extremely diverse in terms of price. They included both low-end products, commonly available in supermarkets, and high-quality, single malt whiskies, including items not available for commercial sale. In such a diverse group of samples, making comparisons analogous to those we made for the samples from Ireland, the USA, or Poland, taking into account the manufacturer, was extremely difficult (more than 60 independent producers were investigated.) There were no statistically significant differences in any case in reference to the brands. Due to the fact that the group of samples from Scotland was much more diversified than those from other countries, the influence of additional parameters, such as the type and age of alcohol, on the grouping of objects was taken into account. Therefore, in the next steps, we verified the hypothesis about the influence of the type of Scotch whisky (blended or single malt) and the aging time (divided into 3 groups based on maturation period) of the single malt whisky from Scotland on the ability to distinguish samples.

3.3. *Elemental Analysis for the Type of Scotch Whisky*

In order to verify the differences in the types of whisky, namely single malt and blended, only products originating from Scotland were taken into account. Therefore, 106 samples were analysed, including 50 single malt whiskies and 56 blended whiskies. The non-parametric test showed the presence of statistically significant differences between the content of such elements as: Al, Cr, Cu, Fe, K, Mg, Mn, P, S, Ti, Tl, Zn, and V (Table S6 and Figure S2A–K). Taking into account the median value for the following elements, in this group, a blended whisky contained more Al, Cr, and Tl when compared with single malt whisky. In turn, for the rest of the elements (Cu, Fe, K, Mg, Mn, P, S, Zn, and V), higher amounts were observed in single malt whisky. Additionally, a projection of the cases on the factor-plane for all samples originating from Scotland was carried out. As shown in the graph, most of the single malt whisky samples were grouped on the left side. In turn, a vast majority of blended whisky samples was placed on the right side of the projection of the cases plot (Figure S3). The outlier point (marked in orange) is a 26-year-old, extremely rare, high-quality single malt whisky, which was not commercially available, and was characterized by an increased content of the following elements: Mn, Co, Cu, Mg, K, and P. The second outlier point (marked in green) was the sample of single malt whisky stored in special, small barrels (octave barrels). Aging the alcohol in much smaller barrels of about 65L will ensure a better integrity of the beverage with the wood. Despite the short maturation period (3 years), the alcohol is much more "saturated" and richer in taste, (in the case of the present study, this sample was characterized by higher levels of the following elements compared to other samples from Scotland: Li, Co, Mo, Bi, and Zn.). In our preliminary studies [12], we performed a semi-quantitative measurement of the following 21 isotopes: ^{44}Ca, ^{45}Sc, ^{47}Ti, ^{48}Ti, ^{51}V, ^{52}Cr, ^{54}Fe, ^{55}Mn, ^{60}Ni, ^{63}Cu, ^{66}Zn, ^{88}Sr, ^{90}Zr, ^{95}Mo, ^{101}Ru, ^{107}Ag, ^{111}Cd, ^{118}Sn, ^{138}Ba, ^{208}Pb, ^{209}Bi, and total Hg content using the ICP-ToF-MS and CV-AAS techniques. We tried to differentiate 20 whisky samples according to country of origin, production region of Scotland, and type of whisky (single malt and blended). The performed analysis revealed the existence of statistically important differences between single malt and blended whiskies for Cr, Fe, Cu, Zn, and Ba. The median counts for copper, chromium, and barium were higher for single malt whisky. In turn, for iron and zinc, the blended whisky samples were characterized by higher counts of the mentioned elements. In our previous study, the analysed samples from various Scottish production regions differed in age, type, and brand. Most of the single malt samples were matured whiskies, while the blended whiskies were mainly 3-year-old products. Unquestionably, this parameter (age) could have affected the existence of statistically significant differences in the whiskies' contents of Cr, Fe, Cu, Zn, and Ba.

Admittedly, the projection of cases on the PCA plot, which was carried out in this study earlier, did not show any grouping by brand or production region. Nevertheless, the obtained PCA graphs potentially suggested a simultaneous overlapping of two parameters, such as age and type. Thus, in order to evaluate the influence of the type of whisky on the elemental compositions, the data set in this work was significantly reduced from 106 objects to 71. Only samples with the same aging period (3 years) were included in the new tested data set, which consisted of 54 blended whiskies and 17 single malt whiskies. In this case, statistically significant differences were reported for such elements as Al, Cr, Cu, Fe, K, Mg, Mn, S, Ti, Tl, and V. Therefore, in relation to the former comparison of all samples from Scotland (106 objects), no statistically significant differences for P and Zn were stated. This allowed us to conclude that these elements (P and Zn) could be related to the age parameter since the influence of this factor was theoretically eliminated as a consequence of the rejection of samples maturated for longer than 3 years.

As the authors of this article showed in the preliminary studies [12] conducted on a much smaller number of objects, the origin of the traces of Cu (Figure 4) could be alembic, which as a rule, is made of copper. This metal enters into a chemical reaction with a distillate and somehow "extracts" sulphuric aromas from it. Moreover, literature reports suggest that copper ions have such a profound effect on the flavour profile of all malt whiskies that has been described as the "fourth ingredient", after malted barley, water, and yeast. Moreover, it was noted that systematic changes within the heating and cooling elements of pot stills can affect copper solubility and hence spirit character [23]. The size of the alembic is extremely important since the longer the distillate touches the copper elements, the softer it will be. Malt whisky is produced in traditional copper stills in batch-type rectors, while grain whisky, which in general contributes the most to the blended whisky, is run continuously using more industrial-style patent stills. Therefore, malt whisky, being distilled in small traditional pot stills, is naturally expected to contain more copper than other types of whisky produced during column still distillation. The sample with the highest Cu content (5252 [μg/L]) was the previously mentioned 26-year-old, single malt whisky. Adam et al., 2002, also confirmed that the whisky had a uniform copper concentration and that the mean copper concentration was significantly higher for all malt whisky samples than for grain and blended scotch whisky samples [10]. The aforementioned grain whisky (with the largest share in blended whisky, especially in the lower price range) is produced with column stills, which are made from stainless steel. This equipment comprises a tall column structure attached above a boiling kettle, and it is designed so as to attain purer vapours [24]. There are several types that are made only of stainless steel, but for the vast majority of them, the composition includes elements such as chromium and nickel. One of the few metals with higher levels in blended whisky was the already-mentioned Cr (from one of the stainless-steel components) (Figure 5). The results obtained for chromium in the blended type of whisky were in the range of 53.83–666.1 [μg/L], whereas for single malt whisky, this range was much narrower (10.70–108 [μg/L]). As mentioned in the introduction, the number of scientific papers on the elemental analysis of whisky, especially those considering its type, is very limited. However, so far, a great deal of work has focused on the analysis of volatile organic compounds eluted using chromatographic techniques [2]. The results show that Scotch grain whisky from a continuous column still distillation contains very few congeners (substances other than the desired type of alcohol, ethanol, produced during fermentation), while Scotch malt whisky produced via double pot still distillation is much richer in them. For the remaining examined elements, despite the lack of statistically significant differences, in most cases, higher concentrations were observed in the single malt whiskies compared to the blended whiskies. Thus, this supports the hypothesis that this type of whisky is richer in various ingredients. The presented results may prove that the equipment used in the alcohol distillation process may have a significant impact on the elemental profile of the final product.

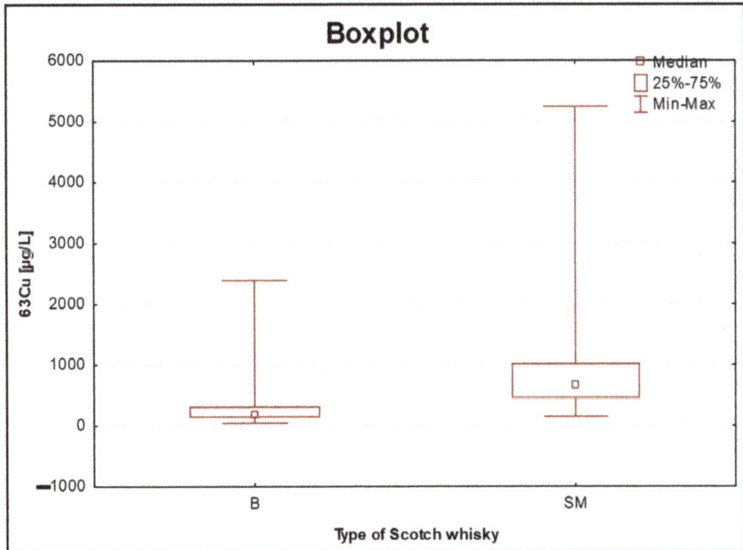

Figure 4. Boxplot for the concentration of Cu obtained for 106 objects of Scotch whisky, divided into two groups: blended (B) and single malt (SM).

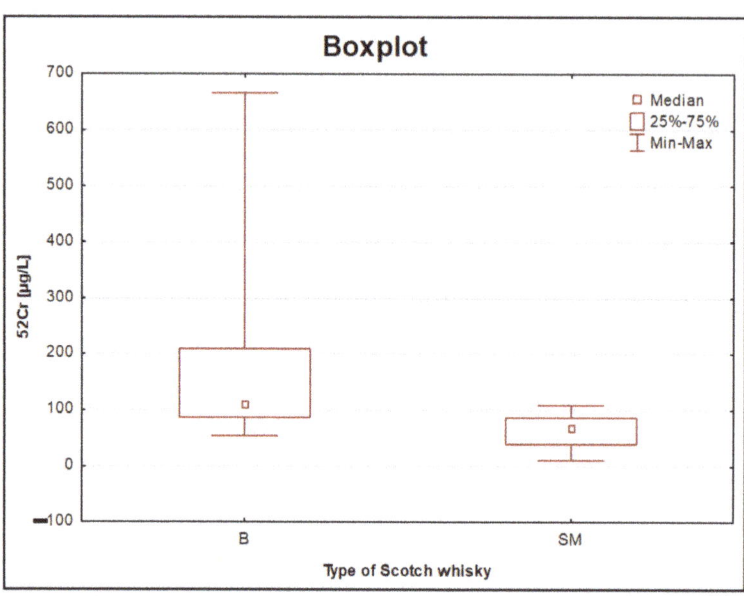

Figure 5. Boxplot for the concentration of Cr obtained for 106 objects of Scotch whisky, divided into two groups: blended (B) and single malt (SM).

3.4. Elemental Analysis for Age of Single Malt Scotch Whisky

In order to verify the hypothesis on the potential impact of aging time on the elemental composition of whisky, from the considered set of samples, 50 objects (Scottish single malt) were selected, where the producer declared the age of the alcohol. The samples were divided into the following groups: 3–9YO; 10–16YO; >16YO (where, in group 3–9 YO, there were only aged products, the minimum period required by law, i.e., 3 years). Considering the tested set of samples in terms of the age of whisky, the existence of statistically significant differences based on Kruskal–Wallis tests only for the concentrations of Cu (3–9YO—10–16YO and 3–9YO—>16YO) and Mn (3–9YO—>16YO) were found (Table 4 and Figure S4A,B in Supplementary Materials). An upward trend was observed for both elements. This means that, the longer the alcohol was aged, i.e., the longer it stayed in the barrel, the higher the content of these elements that was recorded. The mentioned trend is visible when taking into account both the median values and the other basic statistics (minimum and maximum values). It should be emphasized that, in this study, high-quality, single malt Scotch whisky was considered. It was previously proved that this type of alcohol was characterized by a higher content of copper, as a direct consequence of the method of production. The increase in the copper content correlated with the extended aging period is certainly related to the fact that the products with longer aging periods analysed in this study were the leading brands produced in Scottish distilleries. Thus (in accordance with the manufacturers' declarations), dedicated distillation stills with longer "necks" are often used by leading brands in order to ensure special taste qualities. For the rest of the elements determined in this study, again based on the median value in most cases (V, Cr, Ni, Sr, Sb, Bi, Zn, Mg, K, P), despite the lack of statistically significant differences, the same trend was observed as for Cu and Mn. In the previous whisky comparison (single malt and blended), it was concluded that the differentiation of the samples may have been influenced by several, overlapping parameters. Therefore, the authors decided to compare the results only within the group of samples with the same aging period (3 years). Despite the lack of statistically significant differences for the studied elements, the influence of aging on the increased concentrations of Zn and P was clearly visible. Both the mean and median values for these elements increased in the following order: "3–9YO" < "10–16YO" < ">16YO".

Table 4. Contents of selected elements (with statistically significant differences) in the measured Scottish Single Malt Whisky (n = 50) [μg/L].

	Age	n	Mean	Median	Min	Max	Std. Dev.
55Mn	3–9 YO	17	63.09	54.73	16.85	155.1	34.36
	10–16 YO	28	76.93	69.25	22.04	223.1	42.41
	>16 YO	5	133.7	94.17	73.37	260.1	76.67
63Cu	3–9 YO	17	558.2	543.0	143.2	1163	289.9
	10–16 YO	28	982.0	766.6	172.8	2536	606.4
	>16 YO	5	1809	836.7	663.9	5252	1950

As other authors have noted, the concentrations of oak-derived congeners in a given cask of whisky increase with maturation time. Moreover, it is possible to create a graph showing maturation congener concentrations against age. There are literature reports of using near-infrared reflectance (NIR) as a predictive tool for Canadian whisky aging. Natural ^{14}C in atmospheric carbon dioxide is absorbed by metabolism into all plants, including the cereals used for whisky manufacture. Analysis of the ^{14}C levels in ethanol concentrated from the whisky samples was used to estimate the year in which the cereal was grown and then to relate this year with the age of maturation [25]. Chromatographic analysis of selected acids and phenols in chosen samples of whisky (from 6 to 30 YO) conducted by Ng et al., 2000 [26], brought a similar conclusion. The authors stated that, in general, the samples of the oldest whisky contain the highest concentrations of the analysed compounds.

In this work, an interesting relationship between the sulphur concentration and the age of the analysed alcohol beverages samples was made and deserves attention. Sulphur volatile compounds generated during the production process of whisky, to a large degree, influence their quality [27]. On the basis of the research carried out so far, alkyl sulphides such as DMS, DMDS, and DMTS have been recognized as alcohol maturation markers [28,29]. It has been proven that their levels decreased during maturation. In our study, taking into account the median values of S, its levels clearly decreased with age, which undoubtedly had a positive effect on the quality of alcohol. Thus, in our work, the same relationships were proven as made by other authors regarding the S since the sulphur compound levels determined by chromatographic techniques.

Additionally, a projection of the cases on the factor-plane for 50 samples originating from Scotland with the producer's declaration of the age of the alcohol was carried out. Exactly as in the case of the single malt and blended whisky graph, the division of the plot into two parts can be noticed. We can observe a strong tendency that "older" products are on its left side, while the "younger" ones are mostly on the right side of the PCA plot (Figure S5). As for the previous comparison (Figure S3), the same outliers can be identified (i.e., the sample of the unique, 26-year-old whisky marked in orange and the sample of the whisky aged in octave barrels marked in green). Thus, a conclusion can be drawn only about general trends regarding the position of individual samples in the presented projections of cases on the factor-plane, supported by the presence of statistically significant differences. However, it should be emphasized that many parameters affect the possibility of the potential differentiation of particular groups of samples from one another.

4. Conclusions

Taking into account the national standards defining the maximum permissible levels of Cd and Pb in high-percentage alcohol products, it was found that the permissible level was exceeded in the case of Pb for only one sample. The limit value for Cd was not exceeded in any case. For the set of Scotch whisky samples (n = 106), the existence of statistically significant differences was indicated for metals such as Al, Cr, Cu, Fe, K, Mg, Mn, P, S,

Ti, Tl, Zn, and V between the groups of single malt and blended whiskies. Single malt Scotch whisky had a uniform concentration of copper, and the mean copper content was significantly higher for all malt whisky samples than for the blended type. The main source of Cu could be alembic, which, as a rule is made of copper. Moreover, the presented results suggested that the equipment used in the alcohol distillation process may have a significant impact on the elemental profile of the final product.

The analysis of the samples from the USA and from Ireland (n = 26) clearly revealed that the objects that were the same product but originated from independent bottles (e.g., the JB, JDG, and Bus brands) showed similar elemental profiles. From the consumer's point of view, the elemental characteristics of whisky entirely produced in Poland from local raw materials, including home-made products, may seem interesting. Alcohol produced at home can be characterized by the highest content of Sr, K, S, and P as compared to other products from Poland. In terms of the aging time of whisky, the existence of statistically significant differences based on Kruskal–Wallis tests of the concentrations of Cu (3–9YO—10–16YO and 3–9YO—>16YO) and Mn (3–9YO—>16YO) was observed. The conclusion is that the longer the alcohol was aged, i.e., the longer it stayed in the barrel, the higher the content of these elements that was recorded. Based on the comparison of three aging periods only for single malt Scotch whisky, it can be concluded that, despite the lack of statistically significant differences for Zn and P, the influence of aging on the increasing concentration of these elements was clearly visible. Both the mean and median values for these elements increased in the following order: "3–9YO" < "10–16YO" < ">16YO". The study of reduced data set (from 106 to 71 samples) for both type of Scotch whisky samples (single malt and blended) also allowed to conclude that P and Zn could be related to the age parameter. The influence of this factor was theoretically eliminated as a consequence of the rejection of samples matured for no longer than 3 years, no statistically significant differences for these elements were stated. In this study, it has also been proven that levels of sulphur decrease during maturation.

Supplementary Materials: The following supporting information can be downloaded at: https://www.mdpi.com/article/10.3390/foods11111616/s1, Figure S1A–P: Box & whisker plots of selected elements (with statistically significant differences) in the measured whisky samples (n = 170) [μg/L]; Figure S2A–K: Box & whisker plots of selected elements (with statistically significant differences) in the measured Scottish whisky (n = 106) [μg/L]; Figure S3: Projection of the cases on the factor-plane for 106 samples from Scotland according to their type (single malt (SM) and blended (B)); Figure S4A,B: Box & whisker plots of selected elements (with statistically significant differences) in the measured Scottish single malt whisky (n = 50) [μg/L]; Figure S5: Projection of the cases on the factor-plane for 50 samples of single malt whisky from Scotland, Table S1: ICP-MS (Thermo Electron Corporation, X SERIES, East Lyme, CT, USA) and ICP-OES (Thermo Scientific, ICAP 7000 series, Bremen, Germany) parameters and measurement conditions; Table S2: Basic validation parameters obtained for each analyte by using developed method (n, number of standards in three replicates, R^2, coefficient of determination); Table S3: Contents of selected elements (with statistically significant differences) in the measured whisky samples (n = 170) [μg/L]; Table S4: Contents of selected elements (with statistically significant differences) in the measured samples from USA division against the brand (n = 26) [μg/L]; Table S5: Contents of B in the measured samples from Ireland division against the brand (n = 15) [μg/L]; Table S6: Contents of selected elements (with statistically significant differences) in the measured Scottish whisky (n = 106) [μg/L].

Author Contributions: A.P. and M.G. performed an elemental analysis of all samples, analysed the data, performed the chemometric analysis and prepared the paper; K.J. and M.I.S.-J. conducted substantive supervision and final review. All authors have read and agreed to the published version of the manuscript.

Funding: This research received no external funding.

Institutional Review Board Statement: Not applicable.

Informed Consent Statement: Not applicable.

Data Availability Statement: Not applicable.

Acknowledgments: The authors of this paper would like to thank Piotr Wysocki and Patrycja Skrzek for their invaluable technical support.

Conflicts of Interest: The authors declare no conflict of interest.

Sample Availability: Samples are available from the authors.

References

1. Regulation (EC) no 110/2008 of the European Parliament and of the Council of 15 January 2008 on the Definition, Description, Presentation, Labelling and the Protection of Geographical Indications of Spirit Drinks and Repealing Council Regulation (EEC) no 1576/89. Available online: https://eur-lex.europa.eu/legal-content/EN/TXT/PDF/?uri=CELEX:32008R0110&from=EN (accessed on 28 October 2021).
2. Aylott, R. Whisky Analysis. In *Whisky—Technology, Production and Marketing*, 2nd ed.; Russell, I., Stewart, G., Eds.; International Centre for Brewing and Distilling, Heriot-Watt University: Edinburgh, UK, 2014; Volume 14, pp. 243–270.
3. Lyons, T.P. Production of Scotch and Irish whiskies: Their history and evolution. In *The Alcohol Textbook*, 4th ed.; Jacques, K.A., Lyons, T.P., Kelsall, D.R., Eds.; Nottingham University Press: Nottingham, UK, 2003; pp. 193–222.
4. Wiśniewska, P.; Dymerski, T.; Wardecki, W.; Namieśnik, J. Chemical composition analysis and authentication of whisky. *J. Sci. Food Agric.* **2015**, *95*, 2159–2166. [CrossRef] [PubMed]
5. Bendig, P.; Lehnert, K.; Vetter, W. Quantification of bromophenols in Islay whiskies. *J. Agric. Food Chem.* **2014**, *62*, 2767–2771. [CrossRef] [PubMed]
6. Stupak, M.; Goodall, I.; Tomaniova, M.; Pulkrabova, J.; Hajslova, J. A novel approach to assess the quality and authenticity of Scotch Whisky based on a gas chromatography coupled to high resolution mass spectrometry. *Anal. Chim. Acta* **2018**, *1042*, 60–70. [CrossRef] [PubMed]
7. Heather, A.H.; Elkins, J.T. Comparison of unaged and barrel aged whiskies from the same Mash Bill using gas chromatography/mass spectrometry. *J. Brew. Distill.* **2019**, *8*, 1–6. [CrossRef]
8. Daute, M.; Jack, M.; Baxter, I.; Harrison, B.; Grigor, J.; Walker, G. Comparison of Three Approaches to Assess the Flavour Characteristics of Scotch Whisky Spirit. *Appl. Sci.* **2021**, *11*, 1410. [CrossRef]
9. Barbeira, P.J.S.; Stradiotto, N.R. Anodic stripping voltammetric determination of Zn, Pb and Cu traces in whisky samples, Fresen. *J. Anal. Chem.* **1998**, *361*, 507–509.
10. Adam, T.; Duthie, E.; Feldmann, J. Investigations into the use of copper and other metals as indicators for the authenticity of Scotch whiskies. *J. Inst. Brew.* **2002**, *108*, 459–464. [CrossRef]
11. Shand, C.A.; Wendler, R.; Dawson, D.; Yates, K.; Stephenson, H. Multivariate analysis of Scotch whisky by total reflection x-ray fluorescence and chemometric methods: A potential tool in the identification of counterfeits. *Anal. Chim. Acta* **2017**, *976*, 14–24. [CrossRef]
12. Pawlaczyk, A.; Gajek, M.; Jozwik, K.; Szynkowska, M.I. Multielemental Analysis of Various Kinds of Whisky. *Molecules* **2019**, *24*, 1193. [CrossRef]
13. Gajek, M.; Pawlaczyk, A.; Wysocki, P.; Szynkowska-Jozwik, M.I. Elemental Characterization of Ciders and Other Low-Percentage Alcoholic Beverages Available on the Polish Market. *Molecules* **2021**, *26*, 2186. [CrossRef] [PubMed]
14. Konieczka, P.; Namieśnik, J. Walidacja procedur analitycznych. In *Ocena i Kontrola Jakości Wyników Pomiarów Analitycznych*; Konieczka, P., Namieśnik, J., Eds.; Wydawnictwo Naukowo-Techniczne: Warsaw, Poland, 2007; pp. 225–300.
15. Gajek, M.; Pawlaczyk, A.; Szynkowska-Jozwik, M.I. Multi-Elemental Analysis of Wine Samples in Relation to Their Type, Origin, and Grape Variety. *Molecules* **2021**, *26*, 214. [CrossRef] [PubMed]
16. Oiv.int. Available online: http://www.oiv.int/public/medias/3741/e-code-annex-maximumacceptable-limits.pdf (accessed on 13 February 2020).
17. Woldemariam, D.M.; Chandravanshi, B.S. Concentration levels of essential and non-essential elements in selected Ethiopian wines. *Bull. Chem. Soc. Ethiop.* **2011**, *25*, 169–180. [CrossRef]
18. World Health Organization. Guidelines for Drinking-Water Quality, Genewa. 2008. Available online: http://indiawrm.org/HP-2/PDF/waterQualityWHO.pdf (accessed on 15 February 2021).
19. Regulation of the Minister of Health on the Maximum Levels of Biological and Chemical Contaminants that May Be Present in Food, Food Ingredients, Permitted Additives, Processing Aids or on the Surface of Food of 13 January 2003 (Journal of Laws of 2003). Available online: https://www.sejm.gov.pl/ (accessed on 10 December 2021).
20. Lehtonen, P.J.; Keller, L.A.; Ali-Mattila, E.T. Multi-method analysis of matured distilled alcoholic beverages for brand identification. *Z. Lebensm. Forsch. A* **1999**, *208*, 413–417. [CrossRef]
21. González-Arjona, D.; González-Gallero, V.; Pablos, F.; González, A.G. Authentication and differentiation of Irish whiskeys by higher-alcohol congener analysis. *Anal. Chim. Acta* **1999**, *381*, 257–264. [CrossRef]
22. Wiśniewska, P.; Boqué, R.; Borràs, E.; Busto, O.; Wardencki, W.; Namieśnik, J.; Dymerski, T. Authentication of whisky due to its botanical origin and way of production by instrumental analysis and multivariate classification methods. *Spectrochim. Acta Part A Mol. Biomol. Spectrosc.* **2017**, *173*, 849–853. [CrossRef] [PubMed]

23. Bathgate, G.N. The influence of malt and wort processing on spirit character: The lost styles of Scotch malt whisky. *J. Inst. Brew.* **2019**, *125*, 200–213. [CrossRef]
24. Pot Still vs. Column Still. Available online: https://www.winesandspiritsacademy.com/blog/pot-still-vs-column-still/ (accessed on 20 December 2021).
25. Livermore, D. Near infrared reflectance (NIR) used as a predictive tool for Canadian whisky ageing. In *Distilled Spirits Science and Sustainability*, 4th ed.; Walker, G.M., Goodall, I., Fotheringham, F., Murray, D., Eds.; Nottingham University Press: Nottingham, UK, 2012; pp. 35–45.
26. Ng, L.K.; Lafontaine, P.; Harnois, J. Gas chromatographic-mass spectrometric analysis of acids and phenols in distilled alcohol beverages. Application of anion-exchange disk extraction combined with in-vial elution and silylation. *J. Chromatogr. A* **2000**, *873*, 29–38. [CrossRef]
27. Wanikawa, A.; Sugimoto, T. A Narrative Review of Sulfur Compounds in Whisk(e)y. *Molecules* **2022**, *27*, 1672. [CrossRef] [PubMed]
28. Leppänen, O.; Ronkainen, P.; Denslow, J.; Laakso, R.; Lindeman, A.; Nykanen, I. Polysulphides and thiophenes in whisky. In *Flavour of Distilled Beverages: Origin and Development*; Piggott, J.R., Ed.; E. Horwood Ltd.: Chichester, UK, 1983; pp. 206–214.
29. Leppänen, O.; Denslow, J.; Ronkainen, P. A gas chromatographic method for the accurate determination of low concentrations of volatile sulphur compounds in alcoholic beverages. *J. Inst. Brew.* **1979**, *85*, 350–353. [CrossRef]

 foods

MDPI

Communication

Dispersive Liquid-Liquid Microextraction Followed by HS-SPME for the Determination of Flavor Enhancers in Seafood Using GC-MS

Xiaolin Luo, Xiaoyuan Wang, Ming Du and Xianbing Xu *

National Engineering Research Center of Seafood, School of Food Science and Technology, Dalian Polytechnic University, Dalian 116034, China; lxl1617010328@163.com (X.L.); xiaoyuanw122@163.com (X.W.); duming121@163.com (M.D.)

* Correspondence: xianbingxu@dlpu.edu.cn; Tel.: +86-411-86323262

Abstract: The determination of flavor compounds using headspace solid-phase microextraction (HS-SPME) combined with gas chromatography–mass spectrometry (GC-MS) can be severely interfered with by complex food matrices in food systems, especially solid samples. In this study, dispersive liquid-liquid microextraction (DLLME) was applied prior to HS-SPME to efficiently reduce the matrix effect in solid seafood samples. The method had high sensitivity (the quantification limits of maltol and ethyl maltol were 15 and 5 μg/kg, respectively), an excellent linear relationship (R2 ≥ 0.996), and the sample recovery rate was 89.0–118.6%. The relative standard deviation (RSD %) values for maltol and ethyl maltol were lower than 10%. Maltol (from 0.7 to 2.2 μg/g) and ethyl maltol (from 0.9 to 34.7 μg/g) in seafood were detected in the selected samples by the developed method. Finally, DLLME coupled with HS-SPME effectively removed the influence of sample matrix and improved the sensitivity of the method. The developed method was applicable in the analysis of flavor enhancers in complex matrix foods.

Keywords: dispersive liquid-liquid microextraction; HS-SPME; flavor enhancer; seafood; GC-MS

Citation: Luo, X.; Wang, X.; Du, M.; Xu, X. Dispersive Liquid-Liquid Microextraction Followed by HS-SPME for the Determination of Flavor Enhancers in Seafood Using GC-MS. *Foods* **2022**, *11*, 1507. https://doi.org/10.3390/foods11101507

Academic Editor: Daniel Cozzolino

Received: 4 May 2022
Accepted: 20 May 2022
Published: 22 May 2022

1. Introduction

Maltol and ethyl maltol, which are derived from sucrose pyrolysis and food baking, have a caramelized flavor, can enhance the flavor and sweetness of food [1,2], and are often used as food flavor enhancers in seafood processing to cover up a fishy smell and improve the flavor of products [3,4]. However, excessive consumption of flavor enhancers can have adverse effects on health, such as dizziness and nausea [5,6]. Meanwhile, maltol can chelate metal ions to form derivative complexes, which have certain toxicity to cells and affect liver and kidney function [7]. Studies had shown that when maltol was fed to rats at a daily dose of 1000 mg/kg, it caused kidney damage and even death [8]. Therefore, the amount of maltol and ethyl maltol added into the diet should be strictly monitored and an analytical method needs to be established that can improve detection sensitivity, and therefore, accurately detect and control the content of flavor enhancers in food samples.

As compared with the HPLC-MS method [9–11], the GC-MS method, with its specific advantages such as high separation resolution and reliable spectrum library search, has been mainly applied to determinate volatile compounds [12–14]. Headspace solid-phase microextraction (HS-SPME) has been a popular pretreat method applied to enrich volatile compounds for further quantitation and quantification using the GC-MS method [15–17]. Although the headspace method can effectively remove the nonvolatile compounds injected into a GC sample port [18], the enrichment of volatile compounds is severely interfered with by sample matrix, which consequently induces low sensitivity and less robustness for several volatile compounds [19,20]. Therefore, there is an urgent need to reduce the matrix effect of headspace methods for determining volatile compounds using the GC-MS method.

Traditionally, solid phase extraction (SPE) or liquid-liquid extraction (LLE) have been applied to remove most sample matrix [21,22], however, these traditional methods are time-consuming and consume large volumes of solvent [23,24]. As compared with traditional methods, the dispersive liquid-liquid microextraction (DLLME) technique can reduce solvent consumption and can concentrate analytes rapidly, which significantly improves the extraction efficiency [25–27]. In addition, DLLME had been reported to significantly reduce matrix interference in analyses of contaminants in wine [27,28], polycyclic aromatic hydrocarbons (PAHs) in roasted cocoa beans [29], and acrylamide in coffee samples [30]. To the best of our knowledge, there are no reports on HS-SPME combined with DLLME applied to detect flavor enhancers in seafood.

In this study, solid seafood was pretreated using DLLME/HS-SPME, and the flavor substances (maltol and ethyl maltol) in seafood were detected and analyzed by gas chromatography–mass spectrometry. The experimental parameters of DLLME and HS-SPME were optimized, and the accuracy and performance of DLLME/HS-SPME/GC-MS were further evaluated.

2. Materials and Methods

2.1. Chemicals

Maltol (99%) and ethyl maltol (99%) were purchased from Macklin (Shanghai, China). The analytical grade organic reagents dichloromethane (CH_2Cl_2), chloroform ($CHCl_3$), carbon disulfide (CS_2), methanol, acetone, and acetonitrile were from Sigma-Aldrich (St. Louis, MO, USA).

2.2. Preprocessing Method of DLLME/HS-SPME

Four types of seafood such as dried squid, instant squid larvae, instant kelp, and instant small yellow croaker were selected as test samples. Two grams of each sample was mixed with 10 mL of deionized water in a centrifuge tube. In order to extract maltol and ethyl maltol sufficiently from a sample, the mixture was homogenized (6000 rpm) for 3 min with a high-speed homogenizer (XHF-DY, Scientitz, 84 Ningbo, China), and then treated with ultrasound for 10 min (SB-800DT, Ningbo, China). After ultrasonic treatment, the sample aqueous solution was centrifuged (4000 rpm) for 5 min. The supernatant was collected, and then further treated using the DLLME technique.

Food matrix components (such as protein and fatty) interact with organic solvent in DLLME to cause an emulsifying phenomenon, which reduces the volume of the separated organic phase [31]. In DLLME, high preconcentration factor and sufficient volume of precipitated phase must be ensured for further analysis after centrifugation. Generally, the volume of dispersant and extractant are in the ranges of 50–3000 μL and 15–2000 μL, respectively [31]. In this study, extractant (500 μL) and dispersant (1.5 mL) were mixed, and then injected into the sample supernatant (2 mL). After centrifugation (6000 rpm) for 3 min, 200 μL of the lower organic phase mixed with 15 μL of cyclohexanone standard (50 mg/L) was transferred into 20 mL headspace vials and further evaporated under vacuum conditions to eliminate the organic solvent. The dried sample was incubated at a constant temperature, and then further extracted by head-space SPME (DVB/CAR/PDMS, 50/30 m). In this study, the optimal conditions for the DLLME treatment were selected by comparing the effects of different extractants (dichloromethane, chloroform, and carbon disulfide), dispersants (methanol, acetonitrile, and acetone), extractant-to-dispersant volume ratios (1:4, 2:5, and 1:2), and ratios of water sample volume to total volume of dispersant and extractant (2:3, 4:5, and 1:1) on the extraction effect of flavor enhancers. The optimal conditions for the HS-SPME treatment were selected by comparing the results of different incubation temperatures (40 °C, 50 °C, and 60 °C), incubation times (10 min, 15 min, 20 min, and 25 min), and extraction times (10 min, 20 min, 30 min, and 40 min) on flavor enhancer extraction.

2.3. GC-MS Determination of Flavor Enhancers

The flavor substances in the samples were separated using a 7890B series gas chromatograph, and the target substances were quantified using a 5977B series mass spectrometer. Chromatographic conditions of the 7890B were: the chromatographic column was an Agilent 19091S-431UI-5MS capillary column (15 m × 250 μm × 0.25 μm), the sample injection volume was 1 μL, and the injection port temperature was 250 °C. The temperature rise program was 35 °C, containing for 0 min, rising to 220 °C at 5 °C /min, and then, rising to 280 °C at 10 °C /min and containing for 2 min. The carrier gas was high purity helium (99.9%), the flow rate was 1 mL/min, the column pressure was 12.04 psi, and the injection port was in the undivided mode. The GC column was directly connected to an Agilent 5977 B series mass selective detector of ion source for the mass spectrometry analysis. The EI source was used as the ion source, the analyte was ionized in the ion source at 70 eV and 230 °C, and the scanning mass range was 40–400 amu [32].

The matrix effect (ME) in samples was calculated using the following equation:

$$ME\% = \frac{\text{Peak area of standards in matrix} - \text{Peak area of standards in solvent}}{\text{Peak area of standards in solvent}} \times 100\% \quad (1)$$

2.4. Statistical Analysis

The mean and standard deviation of each experiment were calculated. Analysis of variance (ANOVA) was used to determine significant differences ($p < 0.05$) between each experiment using the SPSS software package (IBM SPSS Statistics 20).

3. Results and Discussion

3.1. Optimization of the DLLME/HS-SPME Conditions

To reduce the pretreatment time and to reduce the matrix effect on target compounds, DLLME was used to eliminate matrix interference quickly and to improve the method's sensitivity before the HS-SPME/GC-MS analysis. In this study, the conditions of DLLME were optimized using a single factor design experiment. The maltol and ethyl maltol in the water phase were extracted by organic solvent. As shown in Figure 1a,b, the extraction efficiency was increased when chloroform was performed as an extractant and methanol were performed as a dispersant. Therefore, chloroform and methanol were the best extractant and dispersants for DLLME in this experiment. Chloroform is the most widely used extractant (approximately 42% of published studies use this solvent) [31].

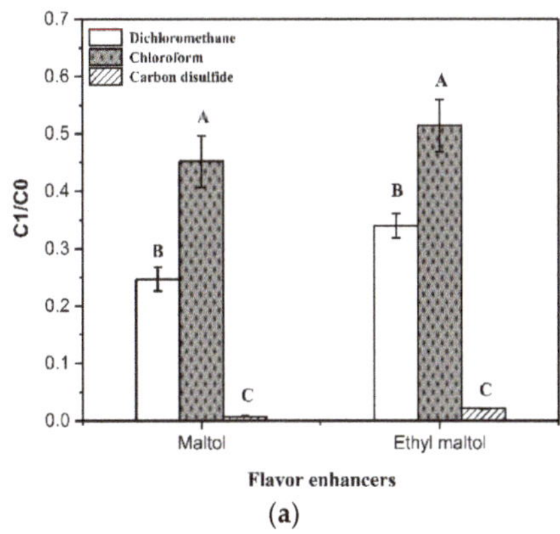

(a)

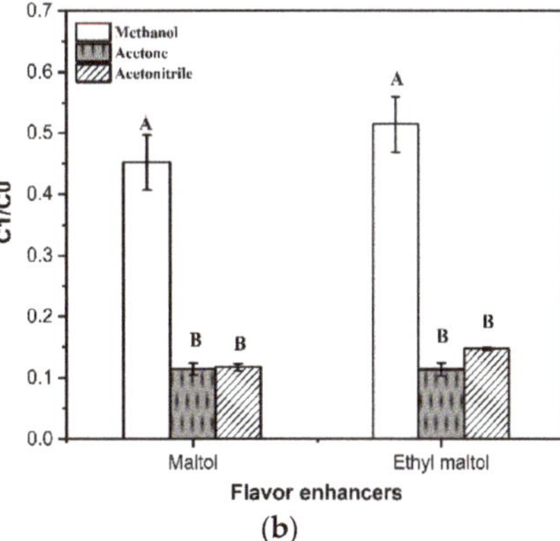

(b)

Figure 1. *Cont.*

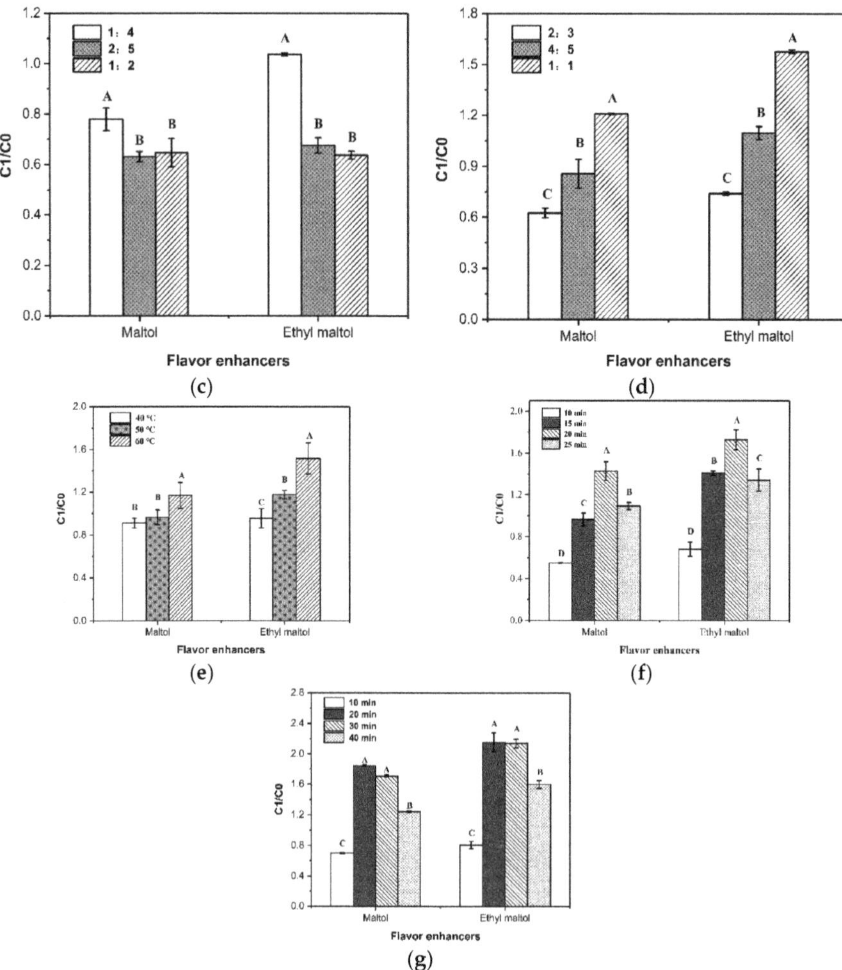

Figure 1. Optimization of conditions for pretreatment dispersive liquid-liquid microextraction/headspace solid-phase microextraction (DLLME/HS-SPME): (**a**) Different types of extractants (dichloromethane, chloroform, and carbon disulfide); (**b**) different types of dispersants (methanol, acetonitrile, and acetone); (**c**) different extractant volume to dispersant volume ratios; (**d**) different aqueous volume to total volume of dispersant and extractant; (**e**) different incubation temperatures (40 °C, 50 °C, and 60 °C); (**f**) different incubation times (10 min, 15 min, 20 min, and 25 min); (**g**) different extraction times (10 min, 20 min, 30 min, and 40 min), on the extraction efficiency of flavor enhancers. Note: C1/C0, the peak area of flavor enhancers (maltol and ethyl maltol)/the peak area of cyclohexanone standard. The superscript letters (A–D) in each histogram indicate significant differences ($p < 0.05$) for the samples.

Similarly, the ratio of each phase in the solution would also affect the extraction efficiency of target analytes. As shown in Figure 1c, the highest extraction efficiency of maltol and ethyl maltol was obtained at 1:4 (the volume ratio of extractant to dispersant). When the volume ratio of extractant to dispersant in the organic phase increased, the extraction efficiency of maltol and ethyl maltol correspondingly decreased due to the dilution effect. In addition, a decrease in organic phase volume ratio could effectively increase the extraction efficiency of maltol and ethyl maltol (Figure 1d). However, the lower volume ratio of organic phase made it difficult to obtain the separated organic phase with enough volume to pipette out. In this study, the volume ratio of the sample (2 mL) to the organic phase was selected as 1:1.

The extraction time, incubating time, and extraction temperature in the HS-SPME method were separately optimized. As shown in Figure 1e, the HS-SPME method was most suitable for removing maltol and ethyl maltol from the DLLME extract when the extraction temperature was 60 °C. Increasing the extraction temperature significantly influenced the extraction of flavor substances. An increase in temperature could increase the content of volatile compounds in the headspace, and therefore, increase extraction efficiency, but excessive high temperature (>60 °C) could induce the desorption of volatile flavor compounds on HS-SPME fiber [32]. Therefore, the best extraction temperature was set at 60 °C. In addition, the incubation time and extraction time had significant effects on the extraction efficiency of HS-SPME. When the incubation time (Figure 1f) and extraction time (Figure 1g) were 20 min separately, the highest extraction efficiency of maltol and ethyl maltol were obtained.

3.2. Verification of Pretreatment Effect of DLLME

In this study, the DLLME/HS-SPME/GC-MS method was evaluated with 1 μg/mL maltol and ethyl maltol standards. As shown in Figure 2a, HS-SPME/GC-MS could effectively detect volatile maltol and ethyl maltol. The chromatographic peaks obtained were well separated (retention time was 13.3 min and 15.8 min, respectively) with symmetrical peak shapes. As compared with the HS-SPME method, the signal intensity of maltol (five times increase) and ethyl maltol (10 times increase) was improved significantly when DLLME was coupled with HS-SPME (Figure 3a).

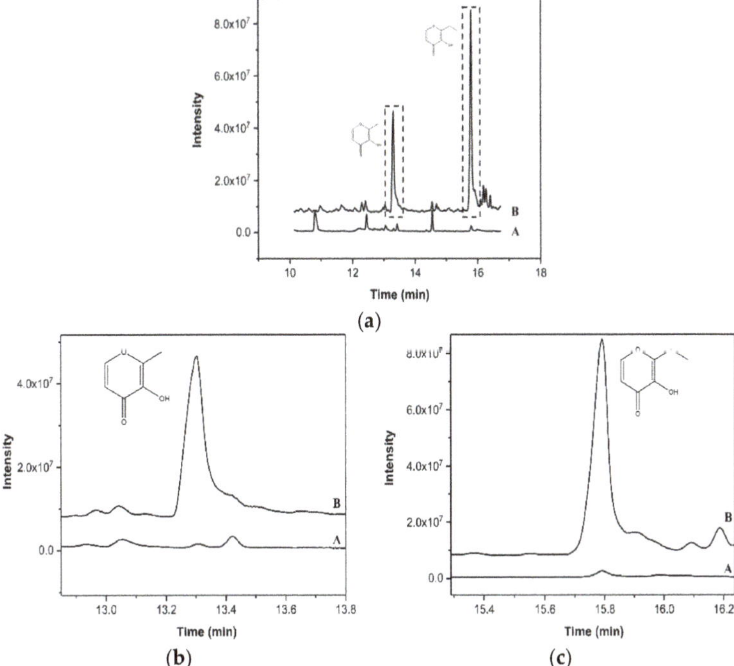

Figure 2. Chromatograms of maltol and ethyl maltol pretreated by dispersive liquid-liquid microextraction combined with headspace solid phase microextraction (A, without DLLME treatment and B, with DLLME treatment): (**a**) Chromatograms of maltol and ethyl maltol standard; (**b**) enlargement of the chromatogram of maltol standard; (**c**) enlargement of the chromatogram of ethyl maltol standard.

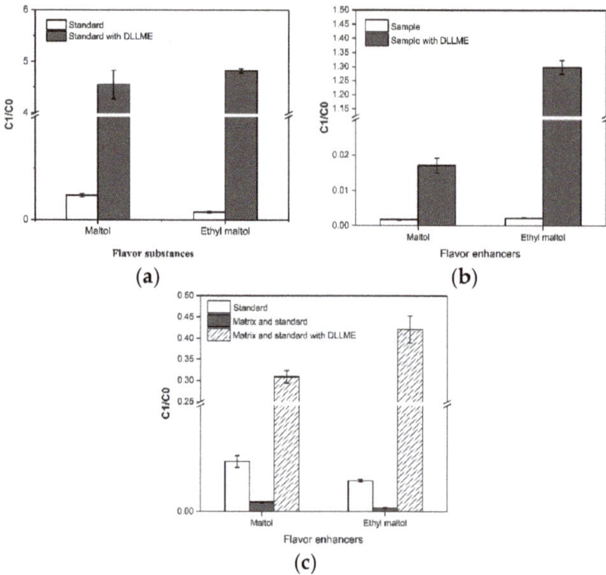

Figure 3. Effects of different pretreatments (without DLLME treatment and after DLLME treatment) on extraction efficiency of maltol and ethyl maltol: (**a**) Effect of DLLME treatment on the extraction efficiency of maltol and ethyl maltol standards; (**b**) effect of DLLME treatment on the extraction efficiency of maltol and ethyl maltol in the matrix (dried squid samples); (**c**) effect of DLLME treatment on the extraction efficiency of maltol and ethyl maltol in the matrix with standards. C1/C0, the peak area of flavor enhancers (maltol and ethyl maltol)/the peak area of cyclohexanone standard.

Complex matrix samples such as seafood contain a high content of oil and protein [33], which significantly interfere with HS-SPME enrichment of maltol and ethyl maltol. Obviously, the signal intensity of maltol and ethyl maltol with sample matrix was individually reduced by five and nine times as compared with the samples without matrix (Figure 3c). However, the signal intensity of maltol and ethyl maltol with matrix was significantly improved after DLLME pretreatment prior to HS-SPME (Figure 3b).

The calculated matrix effect in seafood was −81.12% for maltol and −88.72% for ethyl maltol (Table 1). The enrichment factor of the DLLME method coupled with HS-SPME was 19 for maltol and 66 for ethyl maltol (Table 1). The results showed that DLLME improved the efficiency of the HS-SPME/GC-MS analysis of volatile flavor compounds.

Table 1. Calibration range, limit of detection (LOD), and limit of quantitation (LOQ) for flavor enhancers.

Compounds	Maltol	Ethyl Maltol
Matrix effect (%)	−81.12%	−88.72%
Enrichment factor	19	66
Calibration range (μg/g)	0.25–25.00	0.05–40.00
Regression equation [a]	$y = 0.2806x - 0.0142$	$y = 0.6744x + 0.2082$
R^2	0.9975	0.9967
LOD (μg/kg) [b]	5.0	2.5
LOQ (μg/kg) [c]	15.0	5.0
Intraday precision RSD (n = 3, %)	5.7	2.8
Interday precision RSD (n = 3, %)	4.3	3.5

[a] y is the peak area of flavor substances and x is the concentration of flavor substances; [b] S/N = 3; [c] S/N = 10.

3.3. Evaluation of the Analytical Method

The applicability of DLLME combined with HS-SPME/GC-MS in the analysis of flavor enhancers (maltol and ethyl maltol) in seafood was evaluated based on linear range, sensitivity, stability, and accuracy. The standard curves of maltol and ethyl maltol were obtained by adding maltol standard (0.25–25 μg/g) and ethyl maltol standard (0.05–40 μg/g) into dried squid substrate. As shown in Table 1, the method had an excellent linear relationship, and the correlation coefficients (R^2) of the calibration curves for maltol and ethyl maltol were greater than 0.995. In this study, the limit of detection (LOD) values of maltol and ethyl maltol were 5.0 μg/kg and 2.5 μg/kg, respectively. In addition, the limit of quantification (LOQ) values were 15 μg/kg and 5.0 μg/kg, respectively. As compared with the previously reported methods of solid-phase extraction and dispersed liquid-liquid microextraction (DSPE-DLLME), ionic liquid (IL), and solid-phase extraction (SPE), the LOD and LOQ of the developed method were relatively low (Table 2). Obviously, DLLME coupled with HS-SPME was sensitive for determining maltol and ethyl maltol in complex matrix samples.

Table 2. Comparison of the present method with other methods [a].

Detection Methods	Matrix	Analytes	LOD	LOQ	Reference
DSPE/DLLME-HPLC-PDA	Ready-to-eat seafood	Flavor enhancers (maltol, ethyl maltol, vanillin, methyl vanillin, ethyl vanillin)	60–150 μg/kg	200–500 μg/kg	[34]
IL-IC	Biscuit, chocolate, and milk powder	Spices (vanillin, ethyl vanillin and ethyl maltol)	20–45 μg/kg	70–150 μg/kg	[35]
SPE-GC–MS	Infant formula	Flavoring agents (vanillin, methyl vanillin, ethyl vanillin and coumarin)	-	10 μg/kg	[36]
DLLME/HS-SPME/GC-MS	Seafood	Flavor enhancers (maltol, ethyl maltol)	2.5–5.0 μg/kg	5–15 μg/kg	This work

[a]—not mentioned.

The relative standard deviation (RSD) was detected to evaluate the stability and repeatability of the DLLME/HS-SPME pretreatment method. Intraday precision and interday precision were determined by adding a 5.0 μg/g mixed standard solution of maltol and ethyl maltol to dried squid samples. As shown in Table 1, the intraday accuracy ranged from 2.8 to 5.7% and the interday accuracy ranged from 3.5 to 4.3% for the DLLME/HS-SPME method. The results indicated that the developed method was stable and reliable.

The recovery rate of this method was evaluated by adding three different concentrations of maltol and ethyl maltol standard solutions to dried squid samples. The recovery results showed that the pretreatment of instant dried squid by the DLLME/HS-SPME method did not affect the accuracy of GC-MS detection of flavor substances. As shown in Table 3, the recovery rate of maltol in matrix samples ranged from 89.0 to 118.6%, and that of ethyl maltol in matrix samples ranged from 96.0 to 112.1%. The results suggested the developed method was accurated for determining maltol and ethyl maltol in complex matrix samples.

3.4. Determination of Maltol and Ethyl Maltol in Authentic Seafood

Four kinds of seafood (squid larvae, dried squid, seasoned kelp, and crispy yellow croaker) were pretreated by DLLME, and then further detected and analyzed by HS-SPME/GC-MS. As shown in Table 4, the contents of maltol and ethyl maltol in four kinds of seafood ranged from 0.7 to 2.2 μg/g and from 0.9 to 34.7 μg/g, respectively. The results showed that the method was suitable for the detection and analysis of flavor enhancers in marine products. According to the EU legislation, the added concentration of maltol to food can range from 50 to 200 mg/kg [37]. The recommended maximum acceptable daily dose for the human body is 2 mg/kg [1]. The addition amounts of maltol and ethyl maltol in squid larvae, dried squid, seasoned kelp, and crispy yellow croaker were within the maximum allowable addition range (200 mg/kg).

Table 3. Recovery of maltol and ethyl maltol in samples [a].

Flavor Enhancers	Spiking Level (μg/g)	Recovery (RSD%)
Maltol	1.00	89.0 (4.2)
	12.50	98.1(4.8)
	20.00	118.6 (3.3)
Ethyl maltol	2.50	106.1 (6.8)
	12.50	96.0 (1.1)
	25.00	112.1 (4.4)

[a] Results shown represent % recovery with % RSD in parentheses, $n = 3$.

Table 4. Determination of maltol and ethyl maltol in seafood [a].

Sample	Maltol (μg/g)	Ethyl Maltol (μg/g)
Squid larvae	0.7 (5.5)	1.1 (9.0)
Dried squid	1.4 (8.0)	34.7 (7.5)
Seasoned kelp	2.2 (4.6)	5.3 (3.8)
Crispy yellow croaker	0.7 (8.8)	0.9 (7.1)

[a] % RSD values were given in parentheses, $n = 3$.

4. Conclusions

In this study, flavor enhancers (maltol and ethyl maltol) in solid seafood were detected by using DLLME combined with an HS-SPME/GC-MS analysis. The developed method could significantly eliminate the matrix effect and could significantly improve the method's sensitivity. As expected, DLLME effectively broadened the HS-SPME applications for volatile compounds determination in complex samples.

Author Contributions: Methodology, X.L. and X.W.; validation, X.L. and X.W.; writing—original draft preparation, X.L. and X.X.; investigation, X.W; visualization, X.W.; data curation, X.L.; formal analysis, M.D.; resources, X.L., X.W. and X.X.; project administration, X.L., X.W. and X.X.; funding acquisition, X.X.; conceptualization, X.X.; writing—review and editing, X.X.; supervision, M.D. All authors have read and agreed to the published version of the manuscript.

Funding: This paper is supported by the High-level Talent Innovation Support Program in Dalian, Liaoning Province, China, funded by Dalian Bureau of Science and Technology (grant No. 2019RQ054).

Institutional Review Board Statement: Not applicable.

Informed Consent Statement: Not applicable.

Data Availability Statement: The data that support the findings of this study are available from the corresponding author upon reasonable request.

Acknowledgments: The authors acknowledge Zhenyu Wang, Liming Sun, Liang Dong, and Chao Wu for their considered discussions.

Conflicts of Interest: The authors declare no conflict of interest.

References

1. Alonso, M.; Zamora, L.L.; Calatayud, J.M. Determination of the flavor enhancer maltol through a FIA—Direct chemiluminescence procedure. *Anal. Chim. Acta* **2001**, *438*, 157–163. [CrossRef]
2. Lee, S.J.; Moon, T.W.; Lee, J. Increases of 2-furanmethanol and maltol in Korean red ginseng during explosive puffing process. *J. Food Sci.* **2010**, *75*, C147–C151. [CrossRef] [PubMed]

3. Altunay, N.; Gurkan, R.; Orhan, U. Indirect determination of the flavor enhancer maltol in foods and beverages through flame atomic absorption spectrometry after ultrasound assisted-cloud point extraction. *Food Chem.* **2017**, *235*, 308–317. [CrossRef] [PubMed]
4. Liu, C.; Zhao, L.; Sun, Z.; Cheng, N.; Xue, X.; Wu, L.; Cao, W. Determination of three flavor enhancers using HPLC-ECD and its application in detecting adulteration of honey. *Anal. Methods* **2018**, *10*, 743–748. [CrossRef]
5. Zanchin, G.; Dainese, F.; Trucco, M.; Mainardi, F.; Mampreso, E.; Maggioni, F. Osmophobia in migraine and tension-type headache and its clinical features in patients with migraine. *Cephalalgia* **2007**, *27*, 1061–1068. [CrossRef]
6. Peng, J.; Wei, M.; Hu, Y.; Yang, Y.; Guo, Y.; Zhang, F. Simultaneous Determination of Maltol, Ethyl Maltol, Vanillin, and Ethyl Vanillin in Foods by Isotope Dilution Headspace Solid-Phase Microextraction Coupled with Gas Chromatography-Mass Spectrometry. *Food Anal. Methods* **2019**, *12*, 1725–1735. [CrossRef]
7. Domotor, O.; Aicher, S.; Schmidlehner, M.; Novak, M.S.; Roller, A.; Jakupec, M.A.; Kandioller, W.; Hartinger, C.G.; Keppler, B.K.; Enyedy, E.A. Antitumor pentamethylcyclopentadienyl rhodium complexes of maltol and allomaltol: Synthesis, solution speciation and bioactivity. *J. Inorg. Biochem.* **2014**, *134*, 57–65. [CrossRef]
8. Gralla, E.J.; Stebbins, R.B.; Coleman, G.L.; Delahunt, C.S. Toxicity studies with ethyl maltol. *Toxicol. Appl. Pharmacol.* **1969**, *15*, 604–613. [CrossRef]
9. Risner, C.H.; Kiser, M.J. High-performance liquid chromatography procedure for the determination of flavor enhancers in consumer chocolate products and artificial flavors. *J. Sci. Food Agric.* **2008**, *88*, 1423–1430. [CrossRef]
10. Ma, J.; Zhang, B.; Wang, Y.; Hou, X.; He, L. Determination of flavor enhancers in milk powder by one-step sample preparation and two-dimensional liquid chromatography. *J. Sep. Sci.* **2014**, *37*, 920–926. [CrossRef]
11. Ye, L. Development and validation of a LC-MS/MS method for the determination of isoeugenol in finfish. *Food Chem.* **2017**, *228*, 70–76. [CrossRef] [PubMed]
12. Goncalves, C.; Alpendurada, M.F. Solid-phase micro-extraction-gas chromatography-(tandem) mass spectrometry as a tool for pesticide residue analysis in water samples at high sensitivity and selectivity with confirmation capabilities. *J. Chromatogr. A* **2004**, *1026*, 239–250. [CrossRef] [PubMed]
13. Alder, L.; Greulich, K.; Kempe, G.; Vieth, B. Residue analysis of 500 high priority pesticides: Better by GC-MS or LC-MS/MS? *Mass Spectrom. Rev.* **2010**, *25*, 838–865. [CrossRef] [PubMed]
14. Guerra, P.V.; Yaylayan, V.A. Double Schiff base adducts of 2,3-butanedione with glycine: Formation of pyrazine rings with the participation of amino acid carbon atoms. *J. Agric. Food Chem.* **2012**, *60*, 11440–11445. [CrossRef]
15. Yin, X.; Lv, Y.; Wen, R.; Wang, Y.; Chen, Q.; Kong, B. Characterization of selected Harbin red sausages on the basis of their flavour profiles using HS-SPME-GC/MS combined with electronic nose and electronic tongue. *Meat Sci.* **2021**, *172*, 108345. [CrossRef]
16. Panseri, S.; Soncin, S.; Chiesa, L.M.; Biondi, P.A. A headspace solid-phase microextraction gas-chromatographic mass-spectrometric method (HS-SPME-GC/MS) to quantify hexanal in butter during storage as marker of lipid oxidation. *Food Chem.* **2011**, *127*, 886–889. [CrossRef]
17. Hamm, S.; Bleton, J.; Connan, J.; Tchapla, A. A chemical investigation by headspace SPME and GC-MS of volatile and semi-volatile terpenes in various olibanum samples. *Phytochemistry* **2005**, *66*, 1499–1514. [CrossRef]
18. Ezquerro, Ó.; Pons, B.; Tena, M.A.T. Multiple headspace solid-phase microextraction for the quantitative determination of volatile organic compounds in multilayer packagings. *J. Chromatogr. A* **2003**, *999*, 155–164. [CrossRef]
19. Duran Guerrero, E.; Chinnici, F.; Natali, N.; Marin, R.N.; Riponi, C. Solid-phase extraction method for determination of volatile compounds in traditional balsamic vinegar. *J. Sep. Sci.* **2008**, *31*, 3030–3036. [CrossRef]
20. Wang, X.; Lin, H.; Sui, J.; Cao, L. The effect of fish matrix on the enzyme-linked immunosorbent assay of antibiotics. *J. Sci. Food Agric.* **2013**, *93*, 1603–1609. [CrossRef]
21. Li, J.; Zhang, J.; Liu, Y. Optimization of solid-phase-extraction cleanup and validation of quantitative determination of eugenol in fish samples by gas chromatography-tandem mass spectrometry. *Anal. Bioanal. Chem.* **2015**, *407*, 6563–6568. [CrossRef] [PubMed]
22. Li, J.; Liu, H.; Wang, C.; Wu, L.; Liu, D. Determination of eugenol in fish and shrimp muscle tissue by stable isotope dilution assay and solid-phase extraction coupled gas chromatography-triple quadrupole mass spectrometry. *Anal. Bioanal. Chem.* **2016**, *408*, 6537–6544. [CrossRef] [PubMed]
23. Ouyang, G.; Pawliszyn, J. SPME in environmental analysis. *Anal. Bioanal. Chem.* **2006**, *386*, 1059–1073. [CrossRef]
24. Cortada, C.; Vidal, L.; Canals, A. Determination of nitroaromatic explosives in water samples by direct ultrasound-assisted dispersive liquid-liquid microextraction followed by gas chromatography-mass spectrometry. *Talanta* **2011**, *85*, 2546–2552. [CrossRef]
25. Feng, T.-T.; Liang, X.; Wu, J.-H.; Qin, L.; Tan, M.-Q.; Zhu, B.-W.; Xu, X.-B. Isotope dilution quantification of 5-hydroxymethyl-2-furaldehyde in beverages using vortex-assisted liquid–liquid microextraction coupled with ESI-HPLC-MS/MS. *Anal. Methods* **2017**, *9*, 3839–3844. [CrossRef]
26. Shu, M.; Man, Y.; Ma, H.; Luan, F.; Liu, H.; Gao, Y. Determination of Vanillin in Milk Powder by Capillary Electrophoresis Combined with Dispersive Liquid-Liquid Microextraction. *Food Anal. Methods* **2015**, *9*, 1706–1712. [CrossRef]
27. Pizarro, C.; Sáenz-González, C.; Pérez-Del-Notario, N.; González-Sáiz, J. Development of a dispersive liquid-liquid microextraction method for the simultaneous determination of the main compounds causing cork taint and Brett character in wines using gas chromatography-tandem mass spectrometry. *J. Chromatogr. A* **2011**, *1218*, 1576–1584. [CrossRef] [PubMed]

28. Khalilian, F.; Rezaee, M. Ultrasound-Assisted Extraction Followed by Solid-Phase Extraction Followed by Dispersive Liquid–Liquid Microextraction for the Sensitive Determination of Diazinon and Chlorpyrifos in Rice. *Food Anal. Methods* **2016**, *10*, 885–891. [CrossRef]
29. Agus, B.; Hussain, N.; Selamat, J. Quantification of PAH4 in roasted cocoa beans using QuEChERS and dispersive liquid-liquid micro-extraction (DLLME) coupled with HPLC-FLD. *Food Chem.* **2020**, *303*, 125398. [CrossRef]
30. Galuch, M.B.; Magon, T.; Silveira, R.; Nicácio, A.; Pizzo, J.S.; Bonafe, E.G.; Maldaner, L.; Santos, O.O.; Visentainer, J.V. Determination of acrylamide in brewed coffee by dispersive liquid–liquid microextraction (DLLME) and ultra-performance liquid chromatography tandem mass spectrometry (UPLC-MS/MS). *Food Chem.* **2019**, *282*, 120–126. [CrossRef]
31. Viñas, P.; Campillo, N.; López-García, I.; Hernández-Córdoba, M. Dispersive liquid–liquid microextraction in food analysis. A critical review. *Anal. Bioanal. Chem.* **2014**, *406*, 2067–2099. [CrossRef] [PubMed]
32. Liang, X.; Feng, T.-T.; Wu, J.-H.; Du, M.; Qin, L.; Wang, Z.-Y.; Xu, X.-B. Vortex-Assisted Liquid-Liquid Micro-extraction Followed by Head Space Solid Phase Micro-extraction for the Determination of Eugenol in Fish Using GC-MS. *Food Anal. Methods* **2017**, *11*, 790–796. [CrossRef]
33. Chatterjee, N.S.; Utture, S.; Banerjee, K.; Ahammed Shabeer, T.P.; Kamble, N.; Mathew, S.; Ashok Kumar, K. Multiresidue analysis of multiclass pesticides and polyaromatic hydrocarbons in fatty fish by gas chromatography tandem mass spectrometry and evaluation of matrix effect. *Food Chem.* **2016**, *196*, 1–8. [CrossRef] [PubMed]
34. Ma, Y.J.; Bi, A.Q.; Wang, X.Y.; Qin, L.; Du, M.; Dong, L.; Xu, X.B. Dispersive solid-phase extraction and dispersive liquid-liquid microextraction for the determination of flavor enhancers in ready-to-eat seafood by HPLC-PDA. *Food Chem.* **2020**, *309*, 125753. [CrossRef]
35. Zhu, H.-B.; Fan, Y.-C.; Qian, Y.-L.; Tang, H.-F.; Ruan, Z.; Liu, D.-H.; Wang, H. Determination of spices in food samples by ionic liquid aqueous solution extraction and ion chromatography. *Chin. Chem. Lett.* **2014**, *25*, 465–468. [CrossRef]
36. Shen, Y.; Hu, B.; Chen, X.; Miao, Q.; Wang, C.; Zhu, Z.; Han, C. Determination of four flavorings in infant formula by solid-phase extraction and gas chromatography-tandem mass spectrometry. *J. Agric. Food Chem.* **2014**, *62*, 10881–10888. [CrossRef]
37. Zhou, J.; Zhang, K.; Li, Y.; Li, K.; Ye, B. Study on the electrochemical properties of maltol at a carbon paste electrode and its analytical application. *Anal. Methods* **2012**, *4*, 3206–3211. [CrossRef]

MDPI

Article

Saponification Value of Fats and Oils as Determined from ^{1}H-NMR Data: The Case of Dairy Fats

Mihaela Ivanova [1], Anamaria Hanganu [2,3], Raluca Dumitriu [4], Mihaela Tociu [4], Galin Ivanov [1], Cristina Stavarache [3,5], Liliana Popescu [6], Aliona Ghendov-Mosanu [6], Rodica Sturza [6], Calin Deleanu [3,7] and Nicoleta-Aurelia Chira [4,*]

1 Department of Milk and Dairy Products, Technological Faculty, University of Food Technologies, 26 "Maritsa" Blvd., 4002 Plovdiv, Bulgaria; mivanova@uft-plovdiv.bg (M.I.); ivanovgalin.uft@gmail.com (G.I.)

2 Department of Organic Chemistry, Biochemistry and Catalysis, Research Centre of Applied Organic Chemistry, Faculty of Chemistry, University of Bucharest, 90-92 Panduri Street, 050663 Bucharest, Romania; anamaria_hanganu@yahoo.com

3 "C.D. Nenitescu" Centre of Organic Chemistry of the Romanian Academy, 202B Spl. Independentei, 060023 Bucharest, Romania; crisstavarache@gmail.com (C.S.); calin.deleanu@yahoo.com (C.D.)

4 "C.D. Nenitescu" Organic Chemistry Department, Faculty of Chemical Engineering and Biotechnologies, University POLITEHNICA of Bucharest, 1-7 Polizu Street, 011061 Bucharest, Romania; ralu_dumitriu@yahoo.com (R.D.); mihaela.tociu@upb.ro (M.T.)

5 Advanced Polymer Materials Group, University POLITEHNICA of Bucharest, 1-7 Gh. Polizu Street, 011061 Bucharest, Romania

6 Department of Oenology and Chemistry, Food Technology, Faculty of Food Technology, Technical University of Moldova, 9/9 Studentilor Street, MD-2045 Chisinau, Moldova; liliana.popescu@tpa.utm.md (L.P.); aliona.mosanu@tpa.utm.md (A.G.-M.); rodica.sturza@chim.utm.md (R.S.)

7 "Petru Poni" Institute of Macromolecular Chemistry of the Romanian Academy, Aleea Grigore Ghica Voda 41A, 700487 Iasi, Romania

* Correspondence: nicoleta.chira@chimie.upb.ro

Citation: Ivanova, M.; Hanganu, A.; Dumitriu, R.; Tociu, M.; Ivanov, G.; Stavarache, C.; Popescu, L.; Ghendov-Mosanu, A.; Sturza, R.; Deleanu, C.; et al. Saponification Value of Fats and Oils as Determined from ^{1}H-NMR Data: The Case of Dairy Fats. *Foods* **2022**, *11*, 1466. https://doi.org/10.3390/foods11101466

Academic Editor: Daniel Cozzolino

Received: 14 April 2022
Accepted: 13 May 2022
Published: 18 May 2022

Abstract: The saponification value of fats and oils is one of the most common quality indices, reflecting the mean molecular weight of the constituting triacylglycerols. Proton nuclear magnetic resonance (^{1}H-NMR) spectra of fats and oils display specific resonances for the protons from the structural patterns of the triacylglycerols (i.e., the glycerol backbone), methylene (-CH_2-) groups, double bonds (-CH=CH-) and the terminal methyl (-CH_3) group from the three fatty acyl chains. Consequently, chemometric equations based on the integral values of the ^{1}H-NMR resonances allow for the calculation of the mean molecular weight of triacylglycerol species, leading to the determination of the number of moles of triacylglycerol species *per* 1 g of fat and eventually to the calculation of the saponification value (SV), expressed as mg KOH/g of fat. The algorithm was verified on a series of binary mixtures of tributyrin (TB) and vegetable oils (i.e., soybean and rapeseed oils) in various ratios, ensuring a wide range of SV. Compared to the conventional technique for SV determination (ISO 3657:2013) based on titration, the obtained ^{1}H-NMR-based saponification values differed by a mean percent deviation of 3%, suggesting the new method is a convenient and rapid alternate approach. Moreover, compared to other reported methods of determining the SV from spectroscopic data, this method is not based on regression equations and, consequently, does not require calibration from a database, as the SV is computed directly and independently from the ^{1}H-NMR spectrum of a given oil/fat sample.

Keywords: saponification value; ^{1}H-NMR spectroscopy; tributyrin; dairy fat; vegetable oils

1. Introduction

One of the most common oil quality indices is the saponification value (SV); it is defined as the amount of alkali (expressed as mg KOH/g sample) required to saponify a defined amount of sample. It is conventionally determined through saponification of a known amount of oil/fat with excess KOH solution, followed by back titration of the excess base with acid solution in the presence of phenolphthalein as an indicator. The amount of base needed for saponification of the fatty acyl chains is then indirectly determined from the excess base that remains unreacted. Since the amount (moles) of base reacted is stoichiometrically equal to the amount (moles) of fatty acyl chains contained in 1 g of oil/fat, SV is then dependent on the length of the fatty acyl chains from triacylglycerols. Therefore, a small saponification value indicates long chain fatty acids on the glycerol backbone in a sample; on the contrary, a high SV indicates triacylglycerols with shorter fatty acyl chains. Consequently, SV becomes an easy approach to assess fatty acids' chain length of specific fats/oils.

For example, most of the common oils/fats of vegetable or animal origin (sunflower, soybean, rapeseed, pork lard, beef tallow, chicken fat, etc.) contain almost only long chain fatty acids (C18 and C16), having similar SV values (ranging from 168–196 mg KOH/g oil) [1]. Some vegetable oils, such as the coconut and palm kernel oils, contain large amounts of lauric (C12:0) and myristic (C14:0) acids; therefore, their saponification values are significantly higher (235–260 mg KOH/g oil) [2–5]. Milk fat differs substantially from other fats and oils in terms of the fatty acid profile (FAP), including relevant amounts of short chain (C4–C6) and medium chain (C8–C12) fatty acids, which is subsequently reflected in its high SV (213–227 mg KOH/g fat) [6,7]. Consequently, SV may be helpful in the detection of the adulteration of dairy products with cheaper fats and oils, because the addition of an oil/fat rich in C18 to a dairy product will result in a decrease in the SV.

Although easy and accurate, the reference method of SV determination requires specific glassware and harmful chemicals and is time consuming (according to the protocol, the saponification step takes one hour to complete, because it is critical that the saponification be complete prior to the final titration). In addition, several factors can cause errors in the titration step including misjudging the color of the indicator near the end point, misreading volumes or faulty technique. Therefore, a new, rapid and reliable method would be preferred.

In this respect, spectroscopic methods coupled with multivariate data analysis have attracted attention, being considerably faster and more practical from a procedural viewpoint. For example, SV has been determined through Fourier transform infrared spectroscopy (FTIR) coupled with multivariate analysis [8] with good accuracy, compared to the standard method; however, the main drawback of the methods based on spectroscopic data is that they require the existence of a large spectral base for the model calibration.

^{1}H-NMR spectroscopy is a fast (the recording of a ^{1}H-NMR spectrum takes approximately 2 min) and non-destructive technique that has widely been applied in the analysis of edible oils. ^{1}H-NMR spectra of fats and oils display signals assigned to both the unsaturated moiety and to various methylene groups of the fatty acyl chains. These signals may be used to calculate the average fatty acyl chain length of fat samples. The ^{1}H-NMR technique allows for full process automation, from the recording (due to the autosamplers) to data processing. Small amounts of samples are necessary, which—if needed—can further be recovered simply through solvent evaporation, after the spectra are recorded. Very importantly, the ^{1}H-NMR technique is also reliable, and several papers report the fatty acid profile of fats and oils computed from ^{1}H-NMR data in good agreement with chromatographic data [9–13]. Skiera et al. briefly reported a rapid method for the determination of the SV from NMR data based on the integral of the CH_2 protons adjacent to the ester groups (δ_H 2.2–2.4 ppm) and on the integral of the 1,2,4,5-tetrachloro-3-nitrobenzene (TCNB) signal at δ_H 7.7 ppm, used as an internal standard for quantitative NMR experiments. Five samples (with a single measurement *per* sample) were tested with the new method; the NMR results were in agreement with the values obtained through the ISO method, consequently pointing at the suitability of the NMR spectroscopy for the determination of the quality indices of fats and oils [14].

Based on our previous expertise on NMR chemometrics to edible oils [13], the present work reports a general algorithm for the calculation of the SV of fats and oils from the ^{1}H-

NMR data. The working model consists of a series of binary mixtures of tributyrin (TB) and vegetable oils in various ratios to obtain a wide range of SV. In addition, to ensure an even more variate composition also regarding the unsaturation, soybean and rapeseed oils—SO and RO, respectively—were used to prepare the model samples. The average length of the fatty acyl chains can be computed through chemometric equations from ^{1}H-NMR data, leading to the calculation of the average molecular weight of each sample and eventually to the SV. The new method was evaluated in comparison with the conventional method based on titration and was further applied to a series of edible fats and oils including butter and cheese extracted fats. Compared to other reported methods of determining the SV from spectroscopic data, the proposed method is not based on regression equations and, consequently, does not require calibration from a database. SV may be computed directly and independently from the ^{1}H-NMR spectrum of a given oil/fat sample.

2. Materials and Methods

2.1. Reagents

CH_2Cl_2 (HPLC purity) and anhydrous $MgSO_4$ were from Sigma–Aldrich, as well as tributyrin (97%). The $CDCl_3$ (isotopic purity 99.8%D) was also from Sigma–Aldrich.

2.2. Binary Oil–Tributyrin Mixtures

A series of binary mixtures of tributyrin (TB) and vegetable oils (RO and SO) in various ratios was prepared to obtain a wide range of SVs. Owing to their different fatty acid profiles, SO and RO were chosen as components for binary mixtures to obtain an even more variate composition also with respect to the unsaturation, thus leading to more reliable results. The specific composition of the RO-TB and SO-TB series is given in the Supplementary Table S1.

2.3. Butter and Cheese Samples

Butter (n = 4) and cheese (n = 9) samples of bovine origin were obtained from Romanian, Bulgarian and Moldavian dairy companies. Butter fat (BF) was extracted from butter samples with CH_2Cl_2, dried on anhydrous $MgSO_4$, followed by evaporation of the solvent. Cheese fat was extracted according to ISO 1735 | IDF 5:2004 protocol [15].

2.4. Oil and Fat Samples

Soybean, rapeseed and sunflower seeds were obtained from the National Agricultural Research and Development Institute of Fundulea (NARDI Fundulea), Romania. The oil was extracted from seeds according to the standard Soxhlet protocol [16]. Beef and sheep tallow were extracted with CH_2Cl_2 from subcutaneous adipose tissue, dried on anhydrous $MgSO_4$, followed by evaporation of the solvent. Coconut oil was purchased from Trio Verde S.R.L., Romania (distributor), and the palm stearin and palm kernel oil were from Scintilla Silk, Romania (distributor).

2.5. Saponification Value

The saponification value was determined according to the ISO 3657:2013 standard procedure [17].

2.6. ^{1}H-NMR Spectra

^{1}H-NMR experiments were recorded in a field of 6.9 T using a Bruker Fourier spectrometer (Bruker Biospin, Ettlingen) operating at an ^{1}H Larmor frequency of 300.18 MHz. The ^{1}H-NMR experiments were using the standard zg30 pulse sequence and had the following parameters: 30° pulse, 5.37 s acquisition time, 6.1 kHz spectral window, 16 scans, 65K data points, 1 s delay time; all spectra were recorded at 25 °C. Fat samples (200 mg) were dissolved in 0.6 mL $CDCl_3$ and transferred to 0.5 mm NMR tubes of the type Norell NOR508UP7-5EA (Sigma–Aldrich, Saint Louis, MO, USA). MestReNova 6.0.2-5475 software (Mestrelab Research, Santiago de Compostela, Spain) was used to process the spectra.

To eliminate operator errors, fixed integration limits were used to obtain the integration values (Supplementary Materials Table S1). In addition, for each sample the F resonance (given by the two protons adjacent to the ester group) was considered as a reference and, therefore, calibrated to 2.000; consequently, the rest of the integrals were automatically reported to the reference. According to the general rule for signals integration (i.e., from baseline to baseline), partially overlapping signals were integrated altogether (i.e., A + B and I + J, respectively). The NMR tubes were in-house quality checked as we previously reported [18].

2.7. Statistics

The experiments were run in triplicate (NMR) and in duplicate (ISO 3657:2013). The results are expressed as the mean values ± standard deviation (sd). Tuckey's test was applied for the significantly different means ($p < 0.05$).

3. Results and Discussions

3.1. ^{1}H-NMR Spectral Characterization of Fats and Oils

A typical ^{1}H-NMR spectrum of an oil is illustrated for a rapeseed oil (RO) in Figure 1. The corresponding peak assignment is explained in Table 1. Figure 1 also shows a comparison of the ^{1}H-NMR spectra of tributyrin (TB) and two rapeseed oil–tributyrin binary mixture: RO (30%) + TB (70%) and RO (60%) + TB (40%).

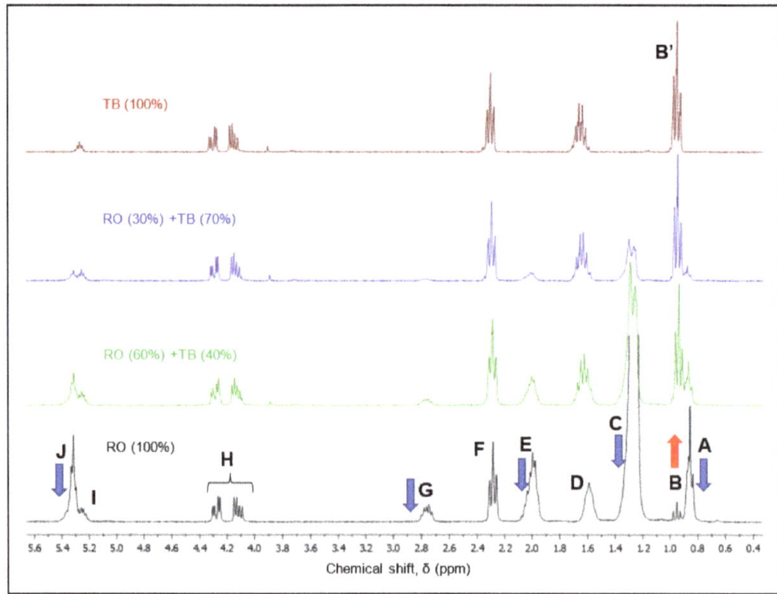

Figure 1. Comparative ^{1}H-NMR spectral characterization of tributyrin (TB —), rapeseed oil (RO —) and rapeseed oil–tributyrin binary mixtures: RO (30%) + TB (70%) — and RO (60%) + TB (40%) —. Letters A–J were assigned to resonances according to letters in Table 1.

Table 1. Chemical shifts and peak assignment of ^{1}H-NMR spectra of milk fats. Adapted with permission from Refs. [12,19]. Copyright 2004, *Eur. J. Lipid Sci. Technol.*; Copyright 2021, *J. Dairy Sci.*

Resonance *	δ (ppm)	Proton	Compound
A	0.85	$-CH_2-CH_2-CH_2-\mathit{CH}_3$	All acids except butyric acid and linolenic acid
B	0.96	$-CH{=}CH-CH_2-\mathit{CH}_3$	Linolenic acid
		$-OOC-CH_2-CH_2-\mathit{CH}_3$	Butyric acid (**B′**)
C	1.24	$-(\mathit{CH}_2)_n-$	All fatty acids
D	1.64	$-\mathit{CH}_2-CH_2-COO-$	All fatty acids
E	2.02	$-\mathit{CH}_2-CH{=}CH-$	All unsaturated fatty acids
F	2.26	$-\mathit{CH}_2-COO-$	All fatty acids
G	2.76	$-CH{=}CH-\mathit{CH}_2-CH{=}CH-$	n-6 (Linoleic) acid and n-3 (linolenic) acid
H	4.19	$-\mathit{CH}_2OCOR$	H in the *sn*-1/3 position of the glycerol backbone
I	5.15	$-\mathit{CH}OCOR$	H in the *sn*-2 position of the glycerol backbone
J	5.29	$-\mathit{CH}{=}\mathit{CH}-$	All unsaturated fatty acids

* Letters from A–J correspond to specific resonances according to Figure 1.

As reflected from Figure 1, certain signals (i.e., A, C, E and J) cannot be found in the spectrum of tributyrin, because butyric acid is a short chain saturated fatty acid, lacking allylic, bis-allylic and unsaturated protons. The butyric moiety displays the triplet B′ characteristic of the terminal methyl group in the structure of fatty acids, the signal D of the protons in position β relative to the ester group, the triplet F generated by the methylene groups adjacent to the ester group and the signals in the specific area of the glycerol backbone (H and I). We have previously shown the assignment of NMR signals in methyl esters of fatty acids as standards for vegetable oil characterization [20]. We have also shown [19] that the resonance characteristic to the terminal methyl group of the fatty acyl chains appears shifted downfield (0.96 ppm) only in the case of linolenic and butyric acyl moieties (B and B′, respectively), compared to the rest of the fatty acyl chains (triplet A, 0.85 ppm). It is therefore evident that as the amount of TB added to the vegetable oil increases, all the resonances related to unsaturated specific groups (J) and those in the vicinity of allylic and bis-allylic groups, (E and G) will decrease. The amplitude of signal C also decreases with the addition of TB, as this resonance is dependent on the length of the fatty acyl chains, being absent for TB.

The only signal that increases in intensity is the triplet B from 0.96 ppm, characteristic for the terminal methyl group in butyric acid or linolenic acid. In rapeseed oil, the 0.96 ppm resonance is due to the linolenic acyl moiety (signal B); as the percentage of added TB increases, this resonance also increases in intensity due to the overlapping signal B′. As expected, the unspecific signals present in all fats and oils, regardless of their specific fatty acid profile (such as H and I from the glycerol moiety, as well as D and F adjacent to the ester group), did not show modifications.

3.2. *Algorithm for the SV Calculation from ^{1}H-NMR Data*

The general pattern of triacylglycerols (TAGs), as depicted in Figure 2, consists of a glycerol ester backbone and three fatty acyl chains, each with a terminal methyl group and various amounts of methylene and CH=CH double bonds.

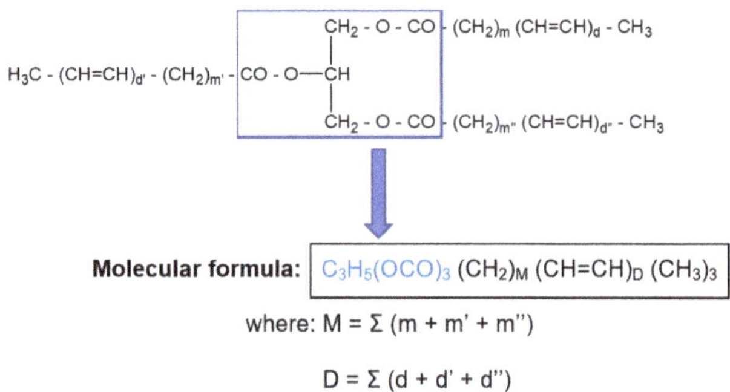

Figure 2. General representation of a triacylglycerol structure.

As reflected from Figure 2, triacylglycerols consist of a glycerol triple ester backbone, common to all TAGs, the differences occurring in the hydrocarbon residues from fatty acyl chains. Apart from the terminal methyl groups (-CH_3), the hydrocarbon chains consist only of methylene groups (-CH_2-) and double bonds (-CH=CH-), the number of which differs depending on the length of the chain and on the degree of unsaturation, being characteristic for each individual fatty acid. For example, oleic acid contains fourteen methylene groups (-CH_2-) and a single double bond (-CH=CH-), and linoleic acid contains twelve methylene groups (-CH_2-) and two double bonds (-CH=CH-). Therefore, the average molecular formula of a triglyceride can be rendered as:

$$C_3H_5(OCO)_3(CH_2)_M(CH{=}CH)_D(CH_3)_3$$

The integral of a resonance being the area under the resonance curve, in the next chemometric equations the following suggestive notations were adopted for the integral values of the corresponding resonances: $A_{(A+B)}$, A_C, A_D, A_E, A_F, A_G, A_H, and $A_{(I+J)}$, respectively.

The average number of methylene groups (M) and the average number of double bonds (D) in the alkyl chain can then be calculated as:

$$M = \frac{3}{2} \cdot \frac{A_C + A_D + A_E + A_F + A_G}{A_{(A+B)}} \quad (1)$$

$$D = \frac{3}{2} \cdot \frac{A_{(I+J)} - A_H/4}{A_{(A+B)}} \quad (2)$$

(i) The normalization factor 3/2 appeared as a consequence of the different number of protons that generated the resonances involved in Equations (1) and (2), i.e., two protons in the case of the resonances at the numerator and three in the case of the resonances at the denominator;
(ii) Since resonances I and J appear partially overlapped, they cannot be integrated separately. However, A_I (corresponding to the single proton in the sn-2 position from the glycerol moiety) can be indirectly computed as $A_H/4$, given the proton ratio of 1:4 in the case of signals I and H, respectively. Consequently, A_J (corresponding to the unsaturated protons (CH=CH) may be computed as a difference $A_{(I+J)} - A_I$;
(iii) Since resonances A and B appear partially overlapped, they cannot be accurately integrated as separate signals; the integration was therefore performed according to the general rule (i.e., from baseline to baseline), leading to the integral of the envelope resonance (A+B).

The mean number of carbon atoms in the hydrocarbon chain (n_C) and the average number of hydrogen atoms in the hydrocarbon chain (n_H) can be computed as:

$$n_C = M + 2D + 1 \quad (3)$$

$$n_H = 2M + 2D + 3 \tag{4}$$

leading to the mean formulae of the hydrocarbon chain ($C_{M+2D+1}H_{2M+2D+3}$) and of the triacylglycerol, i.e., $C_{6+3\,(M+2D+1)}\,H_{5+3(2M+2D+3)}\,O_6$.

As a consequence, the average molecular weight of TAGs becomes:

$$M_{TAG} = 12 \times [6 + 3(M + 2D + 1)] + 1 \times [5 + 3(2M + 2D + 3)] + 16 \times 6 \tag{5}$$

The SV represents the amount of KOH (in mg) required for the saponification of 1 g of fat [15]. Therefore, SV can be computed as:

$$\text{SV (mg KOH/g fat)} = 3 \times \nu \times 56 \times 10^3 \tag{6}$$

where ν represents the number of TAG moles *per* gram of fat ($\nu = 1/M_{TAG}$), while ($3 \times \nu$) is the number of moles of ester groups *per* gram of oil.

An example of SV calculation from ^{1}H-NMR data is shown in the Supplementary Materials (Table S2).

The SV values for the SO-TB and RO-TB series (both determined by the method based on the ^{1}H-NMR data and determined experimentally by the conventional ISO 3657:2013 method taken as reference) are presented in Table 2.

Table 2. SVs determined from the ^{1}H-NMR data and through the standard (i.e., ISO 3657:2013) method for the SO-TB and RO-TB series (95% confidence level).

SO-TB Series				RO-TB Series			
		SV * (mg KOH/g Fat)				SV * (mg KOH/g Fat)	
Sample	TB (%)	From ^{1}H-NMR Data	According to ISO 3657:2013	Sample	TB (%)	From ^{1}H-NMR Data	According to ISO 3657:2013
SO-TB-0	0	196 ± 2 [aA]	190 ± 0 [aB]	RO-TB-0	0	196 ± 4 [aA]	192 ± 1 [aA]
SO-TB-10	10	230 ± 4 [bA]	225 ± 6 [bA]	RO-TB-10	10	233 ± 3 [bA]	227 ± 3 [bA]
SO-TB-20	20	266 ± 2 [cA]	274 ± 3 [cA]	RO-TB-20	20	272 ± 2 [pA]	266 ± 6 [nA]
SO-TB-30	30	302 ± 2 [dA]	294 ± 0 [dB]	RO-TB-30	30	305 ± 4 [dA]	312 ± 10 [lA]
SO-TB-40	40	345 ± 3 [eA]	336 ± 12 [eA]	RO-TB-40	40	341 ± 2 [eA]	334 ± 3 [eA]
SO-TB-50	50	387 ± 2 [fA]	374 ± 10 [fA]	RO-TB-50	50	378 ± 2 [qA]	367 ± 9 [fA]
SO-TB-60	60	412 ± 1 [gA]	403 ± 1 [gB]	RO-TB-60	60	414 ± 3 [gA]	411 ± 1 [gA]
SO-TB-70	70	447 ± 1 [hA]	434 ± 2 [hB]	RO-TB-70	70	448 ± 1 [hA]	433 ± 13 [hA]
SO-TB-80	80	492 ± 2 [iA]	480 ± 3 [iB]	RO-TB-80	80	486 ± 3 [rA]	474 ± 9 [iA]
SO-TB-90	90	535 ± 3 [jA]	530 ± 8 [jB]	RO-TB-90	90	523 ± 2 [sA]	515 ± 0 [mB]
SO-TB-100	100	559 ± 2 [kA]	547 ± 2 [kB]	RO-TB-100	100	560 ± 3 [kA]	551 ± 12 [kA]
SO-TB-15	15	250 ± 3 [lA]	241 ± 3 [bA]	RO-TB-5	5	215 ± 2 [tA]	211 ± 0 [oA]
SO-TB-35	35	326 ± 3 [mA]	318 ± 4 [lA]	RO-TB-25	25	286 ± 2 [uA]	292 ± 3 [dA]
SO-TB-55	55	413 ± 1 [gA]	403 ± 5 [gA]	RO-TB-45	45	359 ± 3 [vA]	350 ± 4 [pA]
SO-TB-75	75	467 ± 3 [nA]	477 ± 4 [iA]	RO-TB-65	65	429 ± 2 [wA]	435 ± 4 [hA]
SO-TB-95	95	540 ± 2 [oA]	527 ± 13 [mA]	RO-TB-85	85	503 ± 3 [xA]	499 ± 1 [qA]

[a–x] Means with different letters within a column are significantly different ($p < 0.05$). [A, B] Means with different letters within a row are significantly different ($p < 0.05$). * Determined in triplicate (NMR method) and in duplicate (ISO method); values are reported as the mean $\pm$ sd.

As reflected in Table 2, the values obtained based on the ^{1}H-NMR data were close to the values determined by the conventional method, which reflects the accuracy of the calculation algorithm.

The accuracy of the new method was assessed by calculating for each sample the SV (NMR) deviation from the SV (ISO), taken as a reference and expressed as percentages relative to the SV (ISO) (see details in Table S3). The mean percent deviation of SV (NMR) from SV (ISO) was found to be 2%, which stands for a robust NMR algorithm. The accuracy of the proposed method was also reflected by the SV (NMR) plotted against the SV (ISO) in Figure 3. The concordance between the values obtained by the NMR method and the titration values is reflected by values close to 1 for both the slope of the trendline (in the case of perfect concordance, tg $\alpha = 1$, corresponding to an angle of 45°) and for the coefficient of correlation R^2. As reflected from Figure 3, values close to 1 were obtained for the two parameters, indicating a good correlation between the two methods.

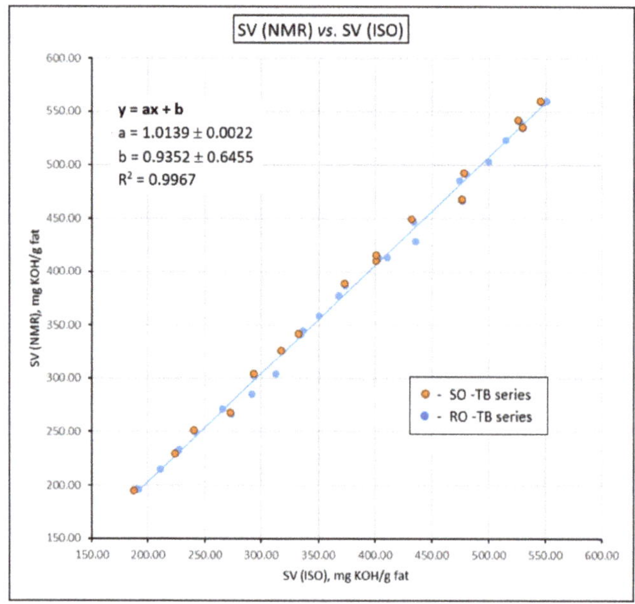

Figure 3. SV (NMR) plotted against the SV (ISO 3657:2013). Values for slope a and intercept b reported as the mean ± sd. The NMR experiments were performed in triplicate; ISO determinations were performed in duplicate.

3.3. Determination of the SV for Edible Oils and Fats

Subsequently, the algorithm for determining the saponification value was applied to a series of commercial samples of vegetable oils and fats, butter, cheeses and spreadable fat mixtures (margarine type). The results are presented in Table 3.

Table 3. SVs determined from ^{1}H-NMR data and through the standard (ISO 3657:2013) method for a series of edible fats and oils (95% confidence level).

No.	Sample	SV * (mg KOH/g Fat)	
		From ^{1}H-NMR Data	According to ISO 3657:2013
	Sunflower oil		
1	Sunflower oil 1	194 ± 2 ^{aA}	188 ± 2 ^{aA}
2	Sunflower oil 2	195 ± 1 ^{aA}	189 ± 2 ^{aA}
3	Sunflower oil 3	194 ± 1 ^{aA}	188 ± 3 ^{aA}
4	Sunflower oil 4	196 ± 1 ^{aA}	188 ± 3 ^{aA}
5	Sunflower oil 5	195 ± 1 ^{aA}	189 ± 2 ^{aA}
6	Rapeseed oil 1	196 ± 1 ^{aA}	188 ± 3 ^{aB}
7	Rapeseed oil 2	196 ± 1 ^{aA}	188 ± 2 ^{aB}

Table 3. *Cont.*

No.	Sample	SV * (mg KOH/g Fat)	
		From ^{1}H-NMR Data	According to ISO 3657:2013
	Rapeseed oil		
8	Rapeseed oil 3	194 ± 1 ^{aA}	188 ± 1 ^{aB}
9	Rapeseed oil 4	195 ± 1 ^{aA}	188 ± 2 ^{aB}
	Soybean oil		
10	Soybean oil 1	195 ± 2 ^{aA}	189 ± 2 ^{aB}
11	Soybean oil 2	193 ± 2 ^{aA}	188 ± 2 ^{aA}
12	Soybean oil 3	194 ± 1 ^{aA}	187 ± 2 ^{aB}
13	Soybean oil 4	195 ± 1 ^{aA}	188 ± 2 ^{aB}
14	Soybean oil 5	194 ± 1 ^{aA}	188 ± 3 ^{aA}
	Coconut oil		
15	Coconut oil 1	249 ± 1 ^{aA}	240 ± 3 ^{aB}
16	Coconut oil 1	248 ± 1 ^{aA}	239 ± 1 ^{aB}
	Palm fat		
17	Palm fat 1	236 ± 1 ^{aA}	230 ± 2 ^{aA}
18	Palm fat 2	237 ± 1 ^{aA}	230 ± 2 ^{aB}
	Butter		
19	Butter 1	242 ± 2 ^{aA}	232 ± 1 ^{aB}
20	Butter 2	245 ± 2 ^{aA}	234 ± 1 ^{aB}
21	Butter 3	245 ± 1 ^{aA}	235 ± 1 ^{aB}
22	Butter 4	239 ± 1 ^{abA}	231 ± 2 ^{aB}
23	Butter 5	241 ± 1 ^{abA}	231 ± 1 ^{aB}
	Spreadable fat mixtures **		
24	Spreadable fat mixture 1	228 ± 1 ^{aA}	217 ± 2 ^{aB}
25	Spreadable fat mixture 2	206 ± 2 ^{bA}	196 ± 1 ^{bB}
26	Spreadable fat mixture 3	222 ± 2 ^{cA}	217 ± 1 ^{aA}
27	Spreadable fat mixture 4	$224 \pm 2a$ ^{acA}	218 ± 1 ^{aB}
	Cheese		
28	Cheese 1	239 ± 2 ^{aA}	231 ± 2 ^{aB}
29	Cheese 2	242 ± 1 ^{aA}	234 ± 1 ^{aB}
30	Cheese 3	244 ± 2 ^{baA}	237 ± 1 ^{baB}
31	Cheese 4	238 ± 1 ^{aA}	231 ± 2 ^{aB}
32	Cheese 5	241 ± 2 ^{aA}	233 ± 3 ^{aA}
33	Cheese 6	241 ± 1 ^{aA}	234 ± 1 ^{aB}
34	Cheese 7	244 ± 2 ^{bA}	237 ± 1 ^{baB}
35	Cheese 8	244 ± 1 ^{bA}	237 ± 2 ^{baB}
36	Cheese 9	239 ± 1 ^{aA}	233 ± 2 ^{aB}

$^{a–c}$ Means with different letters within a column are significantly different ($p < 0.05$). $^{A, B}$ Means with different letters within a row are significantly different ($p < 0.05$). * Determined in triplicate (NMR method) and in duplicate (ISO method), respectively; values reported as the mean ± sd. ** Variable composition (various amounts of butter and different vegetable oils).

As reflected from Table 3, there was agreement between the SVs calculated from the ^{1}H-NMR data and the SVs determined through the wet (ISO 3657:2013) method. However, in the case of the oil and fat samples, the mean percent deviation of SV (NMR) from SV (ISO) was 3%, higher than in the case of the oil–TB series (2%), which may be due to the fact of their more complex composition compared to the binary mixtures.

Edible fats have variable SVs, depending on the species. As expected, vegetable oils, such as sunflower, soybean and rapeseed, had similar SVs, ranging from 194 to 196 mg KOH/g oil (as determined from the ^{1}H-NMR data). These values are in agreement with the fatty acid composition consisting of C18 fatty acids (i.e., linoleic C18:2 and oleic C18:1 as the main constituents, various amounts of stearic C18:0 and linolenic acid C18:3 in small amounts) and modest amounts of C16:0 (palmitic) acid [21–23]. They are also in agreement with similar SVs reported in the literature [21]. On the other hand, lauric fats, such as coconut oil and palm fat, showed significantly higher SVs (mean values of 248.5 and 236.5 mg KOH/g oil, respectively) due to the fact of their specific fatty acid profiles rich in lauric (C12:0), myristic (C14:0) and myristoleic (C14:1) fatty acids. In the case of the coconut oil, its fatty acid profile is dominated by medium chain length fatty acids, with lauric acid ranging between 30 and 50% [24–26], while myristic was also reported in high levels (accounting for more than 20%) [24–26]. Palm fats are abundant in palmitic (C16:0) acid [25,27], with large amounts of lauric and myristic acids (especially palm kernel oil [3]). The high levels of C12 and C14 explain the marked increase in the SVs of coconut and palm fats compared to the rest of the vegetable oils.

In the case of dairy products (i.e., butter and cheese fats), the average saponification values were approximately 242 mg KOH/g fat in both cases. The SV results correlated with their particular fatty acid profile, containing mainly long chain (C14–C18) as well as important amounts of short (butyric, caproic) and medium (C8–C14) chain fatty acids [19,28]. It is worth mentioning that milk fats contain high amounts (up to 32.4% [29]) of palmitic acid (C16:0), whereas myristic (C14:0) and myristoleic (C14:1) acids occur in important amounts, accounting for more than 10–12% altogether [30,31]. Consequently—although belonging to the long chain fatty acids category—C14 fatty acids contributed to the global lowering of the average molecular weight of the triacylglycerols of milk fats compared to vegetable oils (mainly consisting of C16–C18 fatty acids). Altogether, the short and medium chain fatty acids, myristic and palmitic acid levels explain the high SV in the case of dairy products.

On the other hand, spreadable fat mixtures, the analyzed samples consisted of mixtures of butter with various amounts of vegetable fats. Given the variable composition of these samples (depending on the producers' recipes), an average SV cannot be calculated. The spreadable fat mixtures have SV lower than those of butters and cheeses, due to the higher amounts of C16 and C18 fatty acids from the oils and fat ingredients of vegetal origin.

4. Perspectives

Milk fat is one of the most expensive ingredients in the food industry [19,32,33]; therefore, it may be subject to fraudulent practices such as its partial replacement with cheaper oils and fats. The addition of nondairy fats and oils to dairy fats will result in lower SVs. Of course, an altered butter or cheese fat composition would be difficult to detect through SVs if coconut oil (SV = ~249 mg KOH/g oil) combined with a common C16–C18 oil (such as sunflower, rapeseed or soybean oil, with SV = ~193 mg KOH/g oil) is used as an adulterant. On the other hand, except for the producing countries, coconut oil is an expensive commodity [34] in the rest of the regions (for example, in Europe), which makes it improbable as an adulterant. Consequently, SVs may be an indicator for dairy products adulteration with other fats and oils of nondairy origin. Therefore, further studies correlating the amount of vegetable fats added into dairy fats with the variation of the SV may lead to the rapid detection of adulterated dairy products.

5. Conclusions

All structural patterns of triacylglycerols were reflected as specific resonances in the ^{1}H-NMR spectra of fats and oils. Chemometric equations leading to the mean molecular weight of triacylglycerol species may be derived from the integral values of the ^{1}H-NMR signals, which may further be used to compute the number of moles of triacylglycerol species *per* gram of fat, which will further lead to the calculation of the SV, expressed as mg KOH/g of fat. Consequently, ^{1}H-NMR spectroscopic data may be used to rapidly compute the saponification values of oils and fats based on the resonances associated with the fatty acyl chain lengths. The obtained ^{1}H-NMR-based saponification values differed from the conventionally determined SVs by a mean percent deviation of 2.3%, which is sufficient to properly characterize various types of fats. Although the NMR method is more expensive than the official method, as was proven both by us and other groups, one can obtain more information (e.g., fatty acid composition and iodine number) in addition to the saponification value from the same NMR analysis in a very short time. Thus, for combined analyses both for advanced research and authentication purposes, SV by NMR is a valuable alternative.

Supplementary Materials: The following are available online at https://www.mdpi.com/article/10.3390/foods11101466/s1, Examples of the SV algorithm's application—Table S1: Reference intervals for resonance integration; Table S2: 1H-RMN integral values for the UR-TB-20 sample; Table S3: Assessment of the accuracy of the NMR method; Table S4: Influence of various delays on the calculated SV.

Author Contributions: Conceptualization, N.-A.C.; methodology, N.-A.C., C.D. and A.H.; formal analysis, N.-A.C., M.I., R.D. and M.T.; investigation, A.H., R.D., M.I., N.-A.C., M.T., C.S., C.D., L.P. and A.G.-M.; resources, N.-A.C., L.P., M.I., G.I. and R.S.; writing—original draft preparation, N.-A.C. and M.I.; writing—review and editing, C.D., A.H., G.I., R.S. and A.G.-M.; supervision, N.-A.C.; project administration, N.-A.C.; funding acquisition, N.-A.C., M.I. and L.P. All authors have read and agreed to the published version of the manuscript.

Funding: This work was funded through the international research grant "Méthode rapide basée sur la spectroscopie de ^{1}H-RMN pour déceler les fromages adultérés par addition de graisses végétales (FRAUDmage)", code AUF-ECO_SRI_2021_FRAUDmage_2144-2638, financed by the Agence Universitaire de la Francophonie (AUF) and co-funded by the University POLITEHNICA of Bucharest (Bucharest, Romania), the Technical University of Moldova (Chişinău, Republic of Moldova) and the University of Food Technologies (Plovdiv, Bulgaria). The APC was funded by the Agence Universitaire de la Francophonie (AUF) through FRAUDmage research grant.

Institutional Review Board Statement: Not applicable.

Informed Consent Statement: Not applicable.

Data Availability Statement: Data is contained within the article or supplementary material.

Acknowledgments: Support provided by Alina Nicolescu for testing the results on a second NMR spectrometer, as well as occasional support for spectrometers troubleshooting is warmly acknowledged.

Conflicts of Interest: The authors declare no conflict of interest. The funders had no role in the design of the study; in the collection, analyses, or interpretation of data; in the writing of the manuscript, or in the decision to publish the results.

References

1. Li, Y.; Watkins, B.A. Unit D1.4: Oil Quality Indices. In *Current Protocols in Food Analytical Chemistry*; Wrolstad, R.E., Ed.; John Wiley & Sons, Inc.: New York, NY, USA, 2001; Protocol D1.4.3–D1.4.4.
2. Toscano, G.; Riva, G.; Foppa Pedretti, E.; Duca, D. Vegetable oil and fat viscosity forecast models based on iodine number and saponification number. *Biomass Bioenergy* **2012**, *46*, 511–516. [CrossRef]
3. Naksuk, A.; Sabatini, D.A.; Tongcumpou, C. Microemulsion-based palm kernel oil extraction using mixed surfactant solutions. *Ind. Crops Prod.* **2009**, *30*, 194–198. [CrossRef]
4. Kilic, B.; Ozer, C.O. Potential use of interesterified palm kernel oil to replace animal fat in frankfurters. *Meat Sci.* **2019**, *148*, 206–212. [CrossRef] [PubMed]

5. Marina, A.M.; Che Man, Y.B.; Nazimah, S.A.H.; Amin, I. Chemical properties of virgin coconut oil. *J. Am. Oil Chem. Soc.* **2009**, *86*, 301–307. [CrossRef]
6. Sbihi, H.M.; Nehdi, I.A.; Tan, C.P.; Al-Resayes, S.I. Characteristics and fatty acid composition of milk fat from Saudi Aradi goat. *Grassas y Aceites* **2015**, *66*, e101. [CrossRef]
7. Salem, E.R.; Awad, R.A.; El Batawy, O.I. Detection of Milk Fat Adulteration with Coconut Oil Depending on Some Physical and Chemical Properties. *Int. J. Dairy Sci.* **2019**, *14*, 36–44. [CrossRef]
8. Putri, A.R.; Rohman, A.; Setyaningsih, W.; Riyanto, S. Determination of acid, peroxide, and saponification value in patin fish oil by FTIR spectroscopy combined with chemometrics. *Food Res.* **2020**, *4*, 1758–1766. [CrossRef]
9. Alexandri, E.; Ahmed, R.; Siddiqui, H.; Choudhary, M.I.; Tsiafoulis, C.G.; Gerothanassis, I.P. High Resolution NMR Spectroscopy as a Structural and Analytical Tool for Unsaturated Lipids in Solution. *Molecules* **2017**, *22*, 1663. [CrossRef]
10. Yeung, D.K.W.; Lam, S.L.; Griffith, J.F.; Chan, A.B.W.; Chen, Z.; Tsang, P.H.; Leung, P.C. Analysis of bone marrow fatty acid composition using high-resolution proton NMR spectroscopy. *Chem. Phys. Lipids* **2008**, *151*, 103–109. [CrossRef]
11. Siudem, P.; Zielinska, A.; Paradowska, K. Application of ^{1}H NMR in the study of fatty acids composition of vegetable oils. *J. Pharm. Biomed. Anal.* **2022**, *212*, 114658. [CrossRef]
12. Knothe, G.; Kenar, J.A. Determination of the fatty acid profile by ^{1}H-NMR Spectroscopy. *Eur. J. Lipid Sci. Technol.* **2004**, *106*, 88–96. [CrossRef]
13. Chira, N.-A.; Todasca, M.-C.; Nicolescu, A.; Rosu, A.; Nicolae, M.; Rosca, S.-I. Evaluation of the computational methods for determining vegetable oils composition using ^{1}H-NMR spectroscopy. *Rev. Chim.* **2011**, *62*, 42–46. Available online: https://www.revistadechimie.ro/Articles.asp?ID=2863 (accessed on 1 February 2022).
14. Skiera, C.; Steliopoulos, P.; Kuballa, T.; Diehl, B.; Holzgrabe, U. Determination of free fatty acids in pharmaceutical lipids by ^{1}H NMR and comparison with the classical acid value. *J. Pharm. Biomed.* **2014**, *93*, 43–50. [CrossRef] [PubMed]
15. *ISO 1735:2004*; Cheese and Processed Cheese Products—Determination of Fat Content—Gravimetric Method (Reference method). ISO: Geneva, Switzerland, 2004.
16. Shahidi, F. Unit 1.1: Extraction and Measurement of Total Lipids. In *Current Protocols in Food Analytical Chemistry*; Whitaker, J., Ed.; John Wiley & Sons, Inc.: New York, NY, USA, 2001; pp. D1.1.1–D1.1.11. [CrossRef]
17. *ISO 3657:2013*; Animal and Vegetable Fats and Oils—Determination of Saponification Value. ISO: Geneva, Switzerland, 2013.
18. Stavarache, C.; Nicolescu, A.; Duduianu, C.; Ailiesei, G.L.; Balan-Porcarasu, M.; Cristea, M.; Macsim, A.-M.; Popa, O.; Stavarache, C.; Hirtopeanu, A.; et al. A real-life reproducibility assessment for NMR metabolomics. *Diagnostics* **2022**, *12*, 559. [CrossRef]
19. Hanganu, A.; Chira, N.-A. When detection of dairy food fraud fails: An alternative approach through proton nuclear magnetic resonance spectroscopy. *J. Dairy Sci.* **2021**, *104*, 8454–8466. [CrossRef] [PubMed]
20. Deleanu, C.; Enache, C.; Caproiu, M.T.; Cornilescu, G.; Hirtopeanu, A. Esteri metilici ai acizilor grasi. Compusi etalon pentru atributia semnalelor in spectrele RMN de rezolutie inalta ale uleiurilor comestibile. *Rev. Chim.* **1994**, *45*, 1046–1052.
21. Kuang, G.; Du, Y.; Lu, S.; Wang, Z.; Zhang, Z.; Fan, X.; Bilal, M.; Cui, J.; Jia, S. Silica@lipase hybrid biocatalysts with superior activity by mimetic biomineralization in oil/water two-phase system for hydrolysis of soybean oil. *LWT* **2022**, *160*, 113333. [CrossRef]
22. Kampa, J.; Frazier, R.; Rodriguez-Garcia, J. Physical and Chemical Characterisation of Conventional and Nano/Emulsions: Influence of Vegetable Oils from Different Origins. *Foods* **2022**, *11*, 681. [CrossRef]
23. Wang, J.; Han, Y.; Wang, X.; Li, Y.; Wang, S.; Gan, S.; Dong, G.; Chen, X.; Wang, S. Adulteration detection of Qinghai-Tibet Plateau flaxseed oil using HPLC-ELSD profiling of triacylglycerols and chemometrics. *LWT* **2022**, *160*, 113300. [CrossRef]
24. Figueiredo, P.S.; Martins, T.N.; Ravaglia, L.M.; Alcantara, G.B.; Guimarães, R.d.C.A.; Freitas, K.d.C.; Nunes, Â.A.; de Oliveira, L.C.S.; Cortês, M.R.; Michels, F.S.; et al. Linseed, Baru, and Coconut Oils: NMR-Based Metabolomics, Leukocyte Infiltration Potential In Vivo, and Their Oil Characterization. Are There Still Controversies? *Nutrients* **2022**, *14*, 1161. [CrossRef]
25. Dorni, C.; Sharma, P.; Saikia, G.; Longvah, T. Fatty acid profile of edible oils and fats consumed in India. *Food Chem.* **2018**, *238*, 9–15. [CrossRef]
26. Kamath, R.; Basak, S.; Gokhale, J. Recent trends in the development of healthy and functional cheese analogues-a review. *LWT* **2022**, *155*, 112991. [CrossRef]
27. Theam, K.L.; Islam, A.; Choo, Y.M.; Taufiq-Yap, Y.H. Biodiesel from low cost palm stearin using metal doped methoxide solid catalyst. *Ind. Crops Prod.* **2015**, *76*, 281–289. [CrossRef]
28. Faccia, M.; Natrella, G.; Gambacorta, G.; Trani, A. Cheese ripening in nonconventional conditions: A multiparameter study applied to Protected Geographical Indication Canestrato di Moliterno cheese. *J. Dairy Sci.* **2022**, *105*, 140–153. [CrossRef] [PubMed]
29. Salas-Valerio, W.F.; Aykas, D.P.; Hatta Sakoda, B.A.; Ludena-Urquizo, F.E.; Ball, C.; Plans, M.; Rodriguez-Saona, L. In-field screening of trans-fat levels using mid- and near-infrared spectrometers for butters and margarines commercialized in the Peruvian market. *LWT* **2022**, *157*, 113074. [CrossRef]
30. Silva, C.C.G.; Silva, S.P.M.; Prates, J.A.M.; Bessa, R.J.B.; Rosa, H.J.D.; Rego, O.A. Physicochemical traits and sensory quality of commercial butter produced in the Azores. *Int. Dairy J.* **2019**, *88*, 10–17. [CrossRef]
31. Wilms, J.N.; Hare, K.S.; Fischer-Tlustos, A.J.; Vahmani, P.; Dugan, M.E.R.; Leal, L.N.; Steele, M.A. Fatty acid profile characterization in colostrum, transition milk, and mature milk of primi- and multiparous cows during the first week of lactation. *J. Dairy Sci.* **2022**, *105*, 2612–2630. [CrossRef]

32. Oduro, A.F.; Saalia, F.K.; Adjei, M.Y.B. Sensory Acceptability and Proximate Composition of 3-Blend Plant-Based Dairy Alternatives. *Foods* **2021**, *10*, 482. [CrossRef]
33. De Meneses, R.B.; Monteiro, M.L.G.; dos Santos, F.F.; da Rocha-Leão, M.H.M.; Conte-Junior, C.A. Sensory Characteristics of Dairy By-Products as Potential Milk Replacers in Ice Cream. *Sustainability* **2021**, *13*, 1531. [CrossRef]
34. Amit; Jamwal, R.; Kumari, S.; Dhaulaniya, A.S.; Balan, B.; Kelly, S.; Cannavan, A.; Singh, D.K. Utilizing ATR-FTIR spectroscopy combined with multivariate chemometric modelling for the swift detection of mustard oil adulteration in virgin coconut oil. *Vib. Spectrosc.* **2020**, *109*, 103066. [CrossRef]

Article

Spectroscopic and Thermal Characterization of Extra Virgin Olive Oil Adulterated with Edible Oils

Emigdio Chavez-Angel [1,*], Blanca Puertas [2], Martin Kreuzer [3], Robert Soliva Fortuny [4], Ryan C. Ng [1], Alejandro Castro-Alvarez [5,*] and Clivia M. Sotomayor Torres [1,6]

1 Catalan Institute of Nanoscience and Nanotechnology (ICN2), CSIC and BIST, Campus UAB, Bellaterra, 08193 Barcelona, Spain; ryan.ng@icn2.cat (R.C.N.); clivia.sotomayor@icn2.cat (C.M.S.T.)
2 Departamento de Calidad, Döehler Fraga, Member of Döehler Group, Collidors S/N, 22520 Fraga, Spain; blanca.puertaspujadas@doehler.com
3 ALBA Synchrotron Light Source Experiment Division—MIRAS Beamline Cerdanyola del Valles, 08290 Barcelona, Spain; mkreuzer@cells.es
4 Agrotecnio-CeRCA Center, Department of Food Technology, University of Lleida, 25198 Lleida, Spain; robert.soliva@udl.cat
5 Centro de Excelencia en Medicina Traslacional, Laboratorio de Bioproductos Farmacéuticos y Cosméticos, Facultad de Medicina, Universidad de La Frontera, Av. Francisco Salazar 01145, Temuco 4780000, Chile
6 ICREA, Pg. Lluís Companys 23, 08010 Barcelona, Spain
* Correspondence: emigdio.chavez@icn2.cat (E.C.-A.); alejandro.castro.a@ufrontera.cl (A.C.-A.)

Abstract: The substitution of extra virgin olive oil with other edible oils is the primary method for fraud in the olive-oil industry. Developing inexpensive analytical methods for confirming the quality and authenticity of olive oils is a major strategy towards combatting food fraud. Current methods used to detect such adulterations require complicated time- and resource-intensive preparation steps. In this work, a comparative study incorporating Raman and infrared spectroscopies, photoluminescence, and thermal-conductivity measurements of different sets of adulterated olive oils is presented. The potential of each characterization technique to detect traces of adulteration in extra virgin olive oils is evaluated. Concentrations of adulterant on the order of 5% can be detected in the Raman, infrared, and photoluminescence spectra. Small changes in thermal conductivity were also found for varying amounts of adulterants. While each of these techniques may individually be unable to identify impurity adulterants, the combination of these techniques together provides a holistic approach to validate the purity and authenticity of olive oils.

Keywords: edible oils; Raman; photoluminescence; FTIR; thermal conductivity; PCA; 2DCOS

Citation: Chavez-Angel, E.; Puertas, B.; Kreuzer, M.; Soliva Fortuny, R.; Ng, R.C.; Castro-Alvarez, A.; Sotomayor Torres, C.M. Spectroscopic and Thermal Characterization of Extra Virgin Olive Oil Adulterated with Edible Oils. *Foods* **2022**, *11*, 1304. https://doi.org/10.3390/foods11091304

Academic Editor: Daniel Cozzolino

Received: 14 April 2022
Accepted: 28 April 2022
Published: 29 April 2022

Publisher's Note: MDPI stays neutral with regard to jurisdictional claims in published maps and institutional affiliations.

1. Introduction

Olive oil is considered to be one of the best edible oils and an essential component in the Mediterranean diet due to extraordinary organoleptic qualities and a large number of health benefits. According to the *Codex alimentarius* of the Food and Agriculture Organization of the United Nations (FAO) [1], olive oils are classified in three categories: virgin olive, refined olive, and refined olive-pomace oils. These, in turn, are divided into different grades depending on their organoleptic qualities, median defects, and color, among other attributes. From the hierarchy list of grades among these categories, extra virgin olive oil (EVOO) is considered to have the highest nutritional value with various health benefits. Among its nutritional properties, EVOO possesses high antioxidant activity [2,3], exhibits anti-inflammatory effects [3,4], improves the metabolism of carbohydrates in patients with type-2 diabetes [5–8], reduces blood pressure and the risk of hypertension [7,9], and improves vasodilation [10,11], to name a few. These many health benefits have boosted the popularity of olive oil in recent decades [12], although this popularity has also brought about other problems associated with the adulteration and/or deliberate mislabeling of EVOO [13,14]. One of the principal motivations for olive-oil fraud is the large price gap between EVOO and other non-olive oils or even between EVOO and other types of olive oils. Due to its relative scarcity and high production/selling price, unscrupulous processors have been fined for adulterating EVOO with large amounts of cheaper oils. EVOO itself is a vegetal fat with high levels of monounsaturated fatty acids (e.g., 78%) and low levels of saturated acids (e.g., 14%), in contrast to cheap seed oils (e.g., sunflower, corn, and soybean), which have high levels of polyunsaturated fats [1]. Consequently, adulteration with other oils results in the loss of many of the healthy properties of EVOO.

There is a long list of properties that can be tested to ensure the quality of EVOO [15–17]. The standard and official methods to characterize EVOO are gas chromatography (GC) and high-performance liquid-chromatography (HPLC). GC is mainly used to determine the composition of the saponifiable fraction, which contains fatty acids and their derivatives, as well as the unsaponifiable fraction, which contains waxes, aliphatic alcohols, tocopherols, and phenolic compounds, among others. On the other hand, HPLC is mainly used to determine the structure of triglycerides, the quantity of pigments such as chlorophylls and carotenes, and other quality parameters (other than purity). Apart from these official methods, there are a number of alternative and complementary methods that have been suggested over the past decade. Among them, infrared and Raman spectroscopy are gaining attention [18–22].

This work aims at evaluating the ability to detect traces of adulteration in EVOO with three spectroscopic techniques: Raman, photoluminescence (PL), and Fourier-transform infrared (FTIR) spectroscopies. In addition, we explore the use of thermal conductivity as a potential new parameter to be used as a detection tool. Despite its relative measurement simplicity, thermal conductivity has, to date, been overlooked as a figure of merit to determine food purity. The combination of all of these techniques provides an easy method, free of sample pre-processing, to ascertain the quality and authenticity of food.

2. Materials and Methods

Twenty samples of EVOO were intentionally adulterated using five different types of edible oils: sunflower (La Masia, "masiasol", Sevilla, Spain), high oleic (HO) sunflower (Carrefour, "Aceite refinado de girasol", Madrid, Spain), 95–5% soybean–nut blend (La española, "Soy plus", Jaen, Spain), corn (Coosol, "Maiz", Jaen, Spain), and olive-pomace (Carrefour, "Aceite refinado de orujo de oliva", Madrid, Spain), in volume concentrations of 5%, 10%, 20%, and 50%. A single type of EVOO (Salvatge "Les Garrigues", Lleida, Spain) was adulterated, and the oil was provided directly from the factory to guarantee its purity. All samples were stored in a dry place protected from light to preserve their quality (see Figure S1a, Supplementary Materials).

The Raman and photoluminescence spectra were recorded using the same equipment (a T64000 Raman spectrometer using a liquid-nitrogen-cooled Symphony CCD manufactured by HORIBA Jobin Yvon, Chilly-Mazarin, France) optimized in the visible regime (400–800 nm). It was used in single-grating mode with 2400 and 300 lines per mm and a spectral resolution of at least 0.4 cm^{-1} and 0.2 nm for Raman and photoluminescence, respectively. The use of 2400 lines for Raman measurements provides a very high frequency resolution at the cost of a small frequency window. On the other hand, 300 lines allow for a larger spectroscopic window which is ideal for the broad PL signal. The measurements were performed by focusing a diode laser (532 nm) onto a transparent quartz cuvette with a 10× long working distance microscope objective (see Figure S1b,c, Supplementary Materials). The power of the laser was kept as low as possible (<0.5 mW) to avoid any possible damage from self heating of the samples. For the photoluminescence measurements (also known as fluorescence spectroscopy), all samples were measured using 3 accumulations with the same integration time of 0.3 s with a fixed focal plane, to allow for direct comparison between each sample. For each sample, 5 to 10 spectra were recorded at positions on the sample.

FTIR spectra were recorded (64 co-added scans) by a Hyperion 3000 infrared (IR) microscope coupled to a Vertex 70 spectrometer manufactured by Bruker (Billerica, MA,

USA) at the infrared beamline MIRAS of the ALBA synchrotron [23]. Data was recorded with a liquid-nitrogen-cooled MCT detector. A 2–5 μL drop of oil was placed on the center of a piece of ZnSe glass and pressured with a second slide to create a homogenous oil film. The setup was used in the transmission configuration with a spectral resolution of 4 cm^{-1} with a Globar as the infrared light source. The IR light was focused onto the ZnSe slide using a 30× Schwarzschild condenser and collected with a matching objective.

Principal component analysis (PCA) was used to treat the FTIR spectra using the software Orange Data Mining [24]. For each sample, 50 spectra at different sample positions were recorded and concatenated in a large matrix, as displayed in Table S1 in the supplementary information. Prior to the calculation, baseline corrections, spectral normalization, and Savitzky–Golay filters (for smoothing) and derivatives (to reduce scatter effects), were applied to process the spectra (see Figure S2, Supplementary Materials).

The thermal conductivity (k) was determined by the bidirectional three-omega (3ω) method [25,26] over the temperature range T = 298–400 K. The bidirectional 3ω method is based on the measurement of a rise in temperature that is produced by an alternating current (AC) passing through a metallic strip via the Joule heating effect. The metal line is composed of four rectangular pads connected by pins to a narrow wire that is used simultaneously as both a heater and temperature sensor (see Figure S1d, Supplementary Materials). The two outer pads are used to apply the AC current while the inner pads are used to measure the third component voltage (3ω-voltage), which contains the information regarding the temperature rise ΔT. Metal heaters (Cr:Au, 5:95 nm) were deposited by physical vapor deposition onto quartz substrates (5 × 5 × 0.5 mm^3). For the measurement, a drop of oil (~10 μL) was placed on top of the 3ω heater. First, an empty 3ω cell was measured (reference). Then, a second measurement took place using the same cell after the sample to be studied was placed on top of the heater. Assuming that heat transfer occurs only across sample-heater-substrate interfaces, the total measured temperature change of the heater (ΔT_{Total}) is given by [25]:

$$\frac{1}{\Delta T_{Total}} = \frac{1}{\Delta T_{Sample}} + \frac{1}{\Delta T_{Substrate}} \tag{1}$$

where ΔT_{Sample} and $\Delta T_{Sustrate}$ correspond to the temperature fluctuations induced by the sample (oil) and substrate (reference) located at the top and the bottom of the heater, respectively. Lubner et al. [26] showed that the error associated with this interface assumption (Equation (1)) can be as small as 1% if three experimental conditions are fulfilled: (i) the ratio of the thermal diffusivities of the sample (α_{oil}) and the substrate (α_{Sub}), $\alpha_{oil}/\alpha_{Sub} > 10^{-1}$; (ii) the ratio of the thermal conductivities is in the range $10^{-2} < k_{oil}/k_{Substrate} < 1$; and (iii) the excitation frequencies are <100 Hz (low frequency limit). In our case, the room-temperature thermal diffusivity of the oils fluctuated within the range of (0.5–0.8) × 10^{-7} m^2 s^{-1} [27,28], while that of the quartz fluctuated within the range of (0.8–1) × 10^{-7} m^2 s^{-1} [29], i.e., $1 > \alpha_{oil}/\alpha_{Sub} > 0.5$. The k of quartz is ~1.2–1.4 W m^{-1} K^{-1} [29,30], and the k of oils was ~0.15–0.17 W m^{-1} K^{-1} [31,32], i.e., $k_{oil}/k_{Substrate} < 1$. The frequency range used here was (5–100) Hz, which falls within the low frequency limit.

The relationship between the temperature rise and the heat generation rate can be expressed as [33,34]:

$$\Delta T = \frac{P}{l\pi k}\int_0^\infty \frac{\sin^2(xb)}{(xb)^2\sqrt{x^2+iq^2}}dx \tag{2}$$

$$q = \sqrt{4\pi f/\alpha} = \sqrt{4\pi f C_V/k} \tag{3}$$

$$\Delta T = \Delta T_X + i\Delta T_Y \tag{4}$$

where ΔT is the complex temperature rise oscillation; b and l are the heater's half width (5 μm) and length (1 mm), respectively; q is the inverse of thermal penetration depth; C_V is the volumetric heat capacity; $i = \sqrt{-1}$ is the imaginary number; f is the excitation frequency; and P is the AC power. The real and the imaginary parts are proportional to the in-phase ('X') and quadrature ('Y') components of three-omega voltage.

Finally, the thermal conductivity of the oils was found by least square fitting of the in-phase signal using k and C_V as fitting variables. A detailed description of the bidirectional technique and the full development of the equations can be found in the supporting information of our previous works [35,36].

3. Results

3.1. Photoluminescence

The photoluminescence (PL) spectra acquired from pure and adulterated EVOO with different amounts of olive-pomace oil adulterant is depicted in Figure 1a. PL spectra for all of the different adulterated EVOO samples that were adulterated with different oils are depicted in Figure S3a–e, Supplementary Materials. Pure adulterant oils (HO sunflower, sunflower, corn, and soy–nut) do not present any PL activity under this 532 nm excitation. In the case of pure olive-pomace oil, which is a common adulterant oil, the PL is very weak with a clear blue shift in its PL peak relative to that of pure EVOO. Despite the fact that both EVOO and olive-pomace oil are derived from olives, their PL spectra present large differences from one another due to the low concentration of compounds with luminescent properties, such as pigments (e.g., chlorophyll, carotenes, and derivatives), phenols, and tocopherols in the adulterant oils [37]. For EVOO, the strong luminescence around 670 nm and 720 nm is mainly associated to the photosystem of chlorophyll [38]. The first peak is attributed to photosystem I (PSI) and the second peak is due to the combination of photosystems I and II [38]. The strong photoluminescence can be seen even by naked eye (Figure S1b,c, Supplementary Materials).

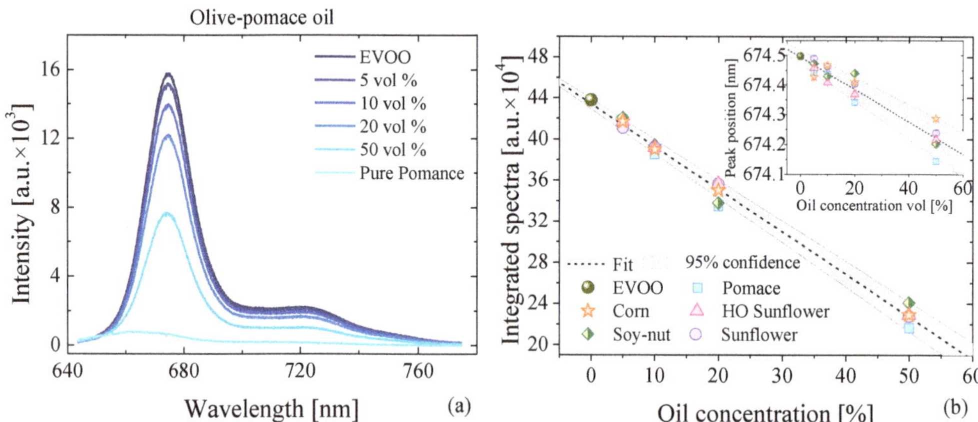

Figure 1. (**a**) Photoluminescence spectra of EVOO adulterated with different concentrations of olive-pomace oil. (**b**) Integrated photoluminescence spectra of the different oil mixtures as a function of the adulterant-oil concentration. The light-grey shaded region represents the 95% range of confidence region around the best-fit line. (Inset) peak position of the PL spectra as a function of adulterant-oil concentration.

The numerically integrated PL intensity of all of the spectra as a function of the adulterant-oil concentration of different adulterant oils is shown in Figure 1b. The light-grey shaded region represents the 95% range of confidence region around the best-fit line. We note that the best-fit line passes through pure EVOO, but a clear linear decrease and variation in the integrated intensity is observed due to the negligible luminescent activity of the adulterant oils. Additionally, we observe a small blue shift of the chlorophyll/pheophytin peak as the concentration of adulterant oil increases (see inset Figure 1b). The origin of this blue shift is even more pronounced when comparing the PL spectra of pure EVOO with pure adulterant oil, since the PL peak in pure adulterant oil sits at lower wavelengths (see Figure S3, Supplementary Materials).

3.2. Raman Spectroscopy

The Raman spectra of pure EVOO, olive-pomace, HO sunflower, corn, and soy–nut blend oils are depicted in Figure 2. Four common bands can be observed in all of the oils, located at ~1265, 1305, 1440 and 1656 cm^{-1}, which correspond to the common Raman modes of unsaturated fatty acids such as: oleic (OA, C18:1), linoleic (LA, C18:2), and linolenic (ALA, C18:3) acid [39]. These molecules are all 18-carbon carboxylic acids with one, two, and three *cis*-double bonds, respectively. Each of the oils under study has a comparable fatty-acid composition (see Table S2 in the supplementary information), which leads to these common carboxylic acid peaks in their Raman spectra. These characteristic Raman bands have already been previously studied by El-Abassy et al. and Lv et al. [39,40]. The attribution of each of the observed peaks to their associated vibrational mode for all spectra in Figure 2 is summarized in Table 1. The remaining two Raman bands located at ~1155 and 1523 cm^{-1}, which are unique to EVOO, have previously been associated with C–C and C=C stretching vibrations of the main polyene chain of carotenoids [41,42]. These additional two bands are not detected in any of the adulterant oils, including the olive-pomace oil. As was the case with the photoluminescence, Raman spectroscopy clearly distinguishes a spectroscopic difference between EVOO and all of the other edible oils, including olive-pomace oil, which shares a common derivation from olives. The absence of carotenoids in refined oils results from the degradation that they suffer during food processing, storage, and thermal treatment. Thermal treatment during the refinement process leads to the isomerization of the carotenoids and a consequent change in their molecular structure [43].

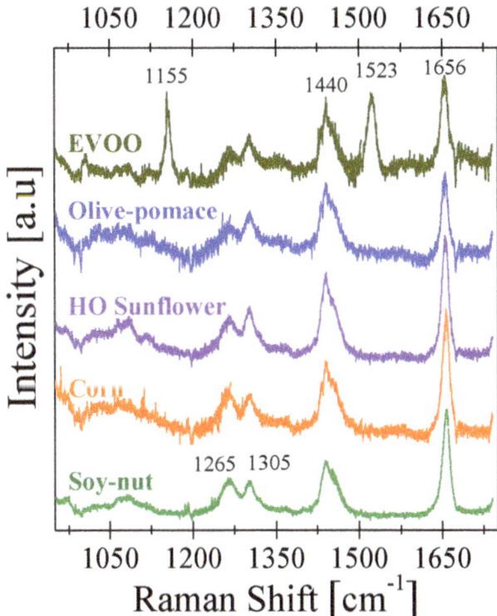

Figure 2. Raman spectrum of pure EVOO, olive-pomace, sunflower, corn, and soy–nut oils.

Table 1. Assignment of the Raman bands of the edible oils.

Frequency [cm^{-1}]	Vibrational Mode
1155	C–C stretching (carotenoid)
1265	=C–H bending scissoring
1305	C–H bending twisting
1440	C–H bending scissoring
1523	C=C stretching (carotenoid)
1656	C=C stretching

Figure 3a shows the normalized Raman spectrum of pure EVOO adulterated with different amounts of HO sunflower. As the HO sunflower concentration increases, the intensity at 1523 cm^{-1} (carotenoid peak) decreases. Similar results were also found for the rest of the adulterant oils (see Figure S4, Supplementary Materials). A summary of these results are shown in Figure 3b, which presents the ratio of the numerical integration of the areas of the Raman peaks at 1523 cm^{-1} and 1656 cm^{-1}. A clear decrease in I_{1523}/I_{1656} ratio can be observed as the adulterant oil content increases. This effect comes from the zero Raman activity for carotenoids peaks shown by all adulterant oils studied here. Notably, the Raman spectra can be directly measured from as-packaged oil without opening and manipulating the oil, allowing for non-invasive verification even from an unopened oil bottle (see Figure S5, Supplementary Materials). Similar to the PL, the addition of adulterant oils leads to a decrease in the I_{1523}/I_{1656} integral ratio. Qui et al. recently observed a similar result using the I_{1523}/I_{1656} ratio to determine the free-fatty-acid (FFA) content of olive oils and found that this intensity ratio decreases linearly with FFA content, although the FFA content was obtained from the nutrition label of each of the oils [21]. Thus, the I_{1523}/I_{1656} integral ratio is an additional useful figure of merit to quantify EVOO purity.

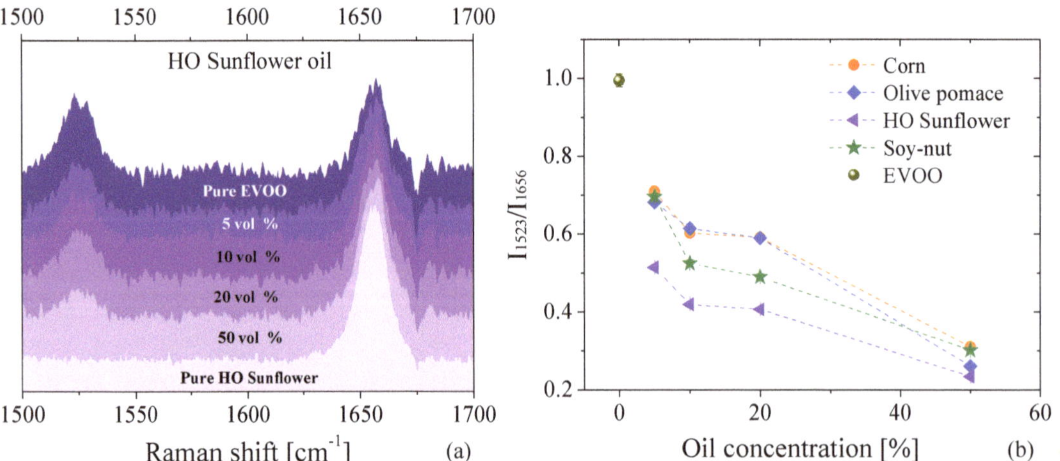

Figure 3. (**a**) Raman spectra of EVOO adulterated with different concentrations of high oleic sunflower oil. (**b**) Intensity ratio of the carotenoid peak (1523 cm^{-1}, I_{1523}) to the C=C stretching peak (1656 cm^{-1}, I_{1656}) as a function of the adulterant oil concentration.

3.3. Fourier-Transform Infrared Spectroscopy

The IR spectra of pure EVOO, corn, soy, and olive-pomace oils are depicted in Figure 4. For ease of visualization, the spectra were separated in two wavenumber ranges: (3150–2800) cm^{-1} and (1500–1000) cm^{-1}. The first window shows the characteristic IR peaks resulting from hydrogen stretching functional groups, while the second window shows other bond deformations and bending that are primarily associated with vibrations of CH_i groups (with $i = 1, 2, 3$) and C–O bonds [44]. Unlike the PL and Raman results, the FTIR spectra shows remarkable similarities between the spectra of the studied samples, making them difficult to differentiate. Therefore, a deep analysis using principal component analysis (PCA), a technique that allows for patterns and variations within a dataset to be readily visualized, was performed to allow for facile differentiation of each of the spectrum from one another. PCA analysis is relatively common in food chemistry, as optical spectra tend to be very similar within particular foods and their associated derivatives. The results of this analysis are displayed in the inset of Figure 4a,b. Our results showed that EVOO and olive-pomace oils could not be differentiated from one another in FTIR spectroscopy, as the PCA scores were almost identical. However, the PCA scores of corn, soy–nut oil, and sunflower oils showed clear differences when compared with EVOO.

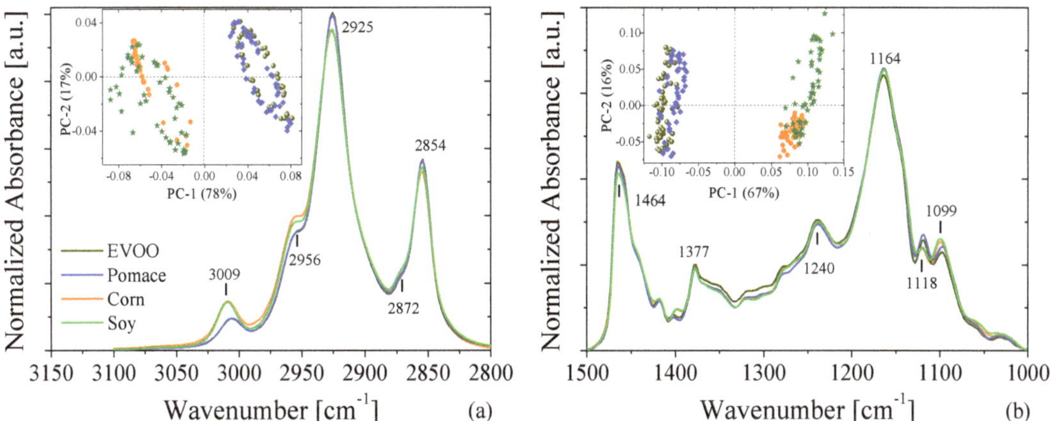

Figure 4. FTIR spectra of four oils: EVOO (pale green), olive-pomace (blue), corn (orange) and soy (dark green) over two different wavenumber ranges focusing on the (**a**) hydrogen stretching functional groups and (**b**) CH_i (i = 1, 2, 3) functional groups present in each oil.

At the most superficial level, a quick differentiation of the IR spectra of the oils was established via PCA, though a deeper analysis of the PCA scores of the adulterated EVOO is possible. Figure 5 shows the subsequent PCA analysis of IR spectra of adulterated EVOO. This rapid and simple PCA analysis highlights the impurities added to EVOO by showing a shift in the scores of adulterated samples as the adulterant oil increases. The shift is observed even with less than 5% of added adulterant (Figure 5a,b,d,e). Similar results have been observed by Vanstone et al. [45], who demonstrated the potential of a combination of near-infrared spectroscopy with PCA to detect EVOO adulteration at levels as low as 2.7%, given an unadulterated reference sample (i.e., pure EVOO). However, we demonstrate similar conclusions with FTIR in the mid-IR spectral range, which is advantageous as molecular fundamental vibrational modes lie in the mid-IR, while spectral measurements in the near-IR are measurements of molecular vibrational overtones. While the PCA alone exhibits potential in its ability to discriminate similar spectra, the addition of a multivariable regression model will be necessary to obtain true quantification of EVOO adulteration.

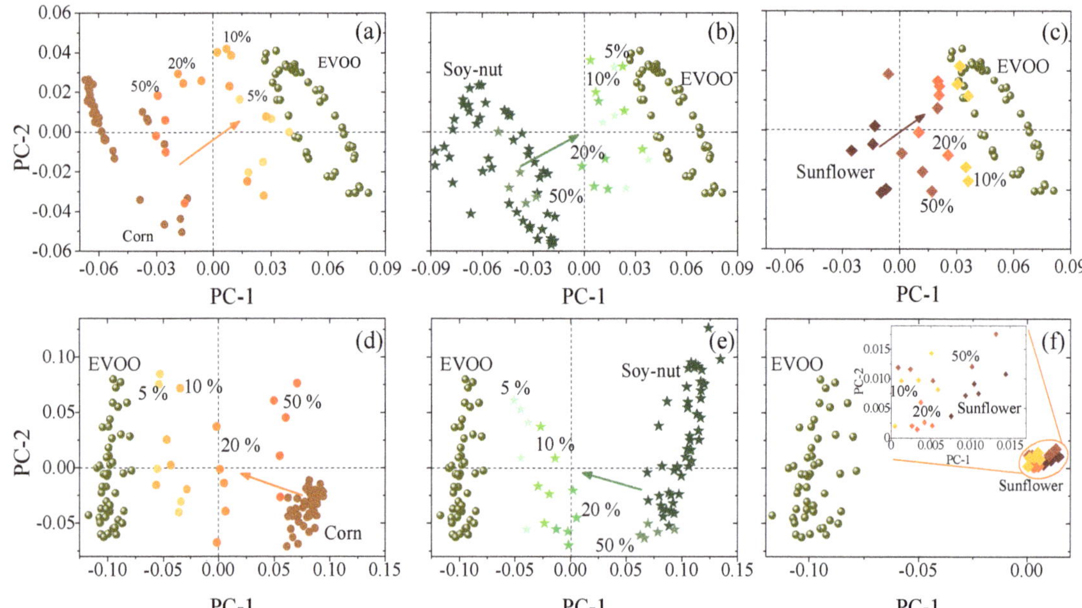

Figure 5. PCA score plots of oil mixtures at the 3000 cm^{-1} and 1300 cm^{-1} window: (**a**,**d**) corn–EVOO, (**b**,**e**) soy-nut–EVOO, and (**c**,**f**) sunflower–EVOO. The inset in (**f**) shows a zoom around the PCA scores of the sunflower-based sample. The color gradient in each figure indicates the evolution of the PCs from pure adulterant (darker colors) to smallest amount of adulterant (lighter colors).

As the PCA of FTIR spectra did not show significant differences between EVOO and olive-pomace, we applied two-dimensional correlation spectroscopy (2DCOS) to gain greater insight into the FTIR spectra. The 2DCOS technique is a mathematical method for analyzing changes in a signal produced by an external perturbation (e.g., a change in temperature, pressure, pH, concentration of mixtures, etc.). To calculate the 2DCOS map we used the concentration of olive-pomace oil as an external perturbation and the spectra dataset was ordered from pure EVOO (0% of oil adulterant) to pure pomace (100% of oil adulterant), i.e., 0, 5, 10, 20, 50, 100%. The raw spectra were baselined and normalized using the most-intense band for each frequency window in Figure 4. The average spectrum was used as a reference spectrum following the same procedure as reference [46]. The 2DCOS analysis was performed with the Mat2dcorr Matlab toolbox [47]. Figure 6 shows the synchronous 2DCOS map at the 3000 cm^{-1} (Figure 6a) and 1300 cm^{-1} (Figure 6b) FTIR frequency windows. The respective autocorrelation and FTIR-averaged spectra are shown above each frequency window. Autocorrelation spectra are defined by a diagonal line along the 2DCOS map and their bands are known as autopeaks. The autopeaks represent real changes between the FTIR spectra that are produced by the external perturbation (such as the addition of olive-pomace oil, in this case). The autocorrelation spectra show three autopeaks located at 2844, 2900, and 2974 cm^{-1} in the 3000 cm^{-1} window and only one autopeak around 1330 cm^{-1} for the 1300 cm^{-1} window. The comparison of the auto correlation and FTIR spectra show that the larger changes among the spectra occur at wavelengths where the FTIR spectra is very weak, which indicates why the PCA analysis was not able to find significant differences between the olive-pomace oil and EVOO. Furthermore, a 2DCOS analysis of pure EVOO oils was also performed using the same data treatment (Figure S6, Supplementary Materials) to verify that the observed variations are not artificial variations resulting from the data treatment such as background subtraction and/or normalization. In this analysis, the same large variation in autopeaks are in fact observed around 2844 and 2900 cm^{-1}, indicating that such variations are dependent not only on real significant differences in oil concentration, but also on experimental fluctuations. Notably, autopeaks around 2974 cm^{-1} are only present in the EVOO/olive-pomace 2DCOS map. Consequently, while this type of data processing enables even such small fluctuations to be used as identifiers for authentication between oils of similar origin, additional processing and identification of 2DCOS peaks may first be required.

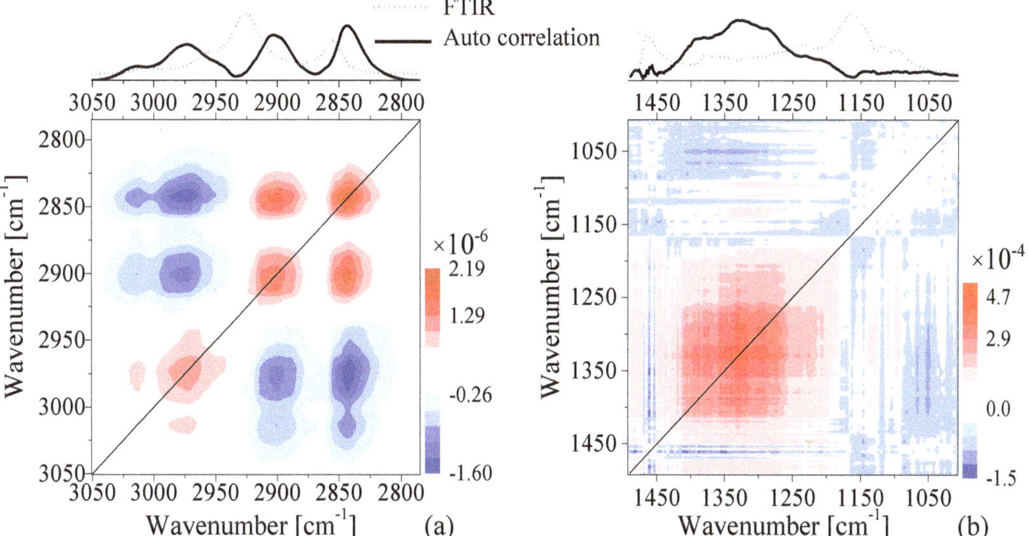

Figure 6. Contour map of the synchronous 2D FTIR correlation spectra of the EVOO–pomace mixtures at the (**a**) 3000 cm^{-1} and (**b**) 1300 cm^{-1} frequency windows. The spectra above the 2D plots provide the auto correlation spectrum (black solid lines) of each 2DCOS map. The average of the FTIR spectra in each window is also included for comparison (grey dashed lines).

3.4. Thermal Conductivities

The temperature dependence of the thermal conductivity (k) of pure oils and of three adulterated mixtures are shown in Figure 7a,b, respectively. A monotonic decrease in the thermal conductivity as the temperature increases can be observed in all of the studied samples. A similar temperature dependence was also reported by Turgut et al. [31]. Interestingly, the temperature dependence of the k of the EVOO–soy–nut oil mixture follows the same temperature dependence as the pure EVOO, even at 50% of soy–nut oil concentration. A comparative analysis of the k values at room temperature (Figure 7c) and at 400 K (Figure 7d) demonstrate noticeable differences between each pure oil and between EVOO and its adulterated mixtures. This highlights the ability of the k—which, to date, has tended to be overlooked as a useful figure of merit in food authentication—to provide information that enables the distinction of pure and adulterated oils, or, more generally, other food products as well.

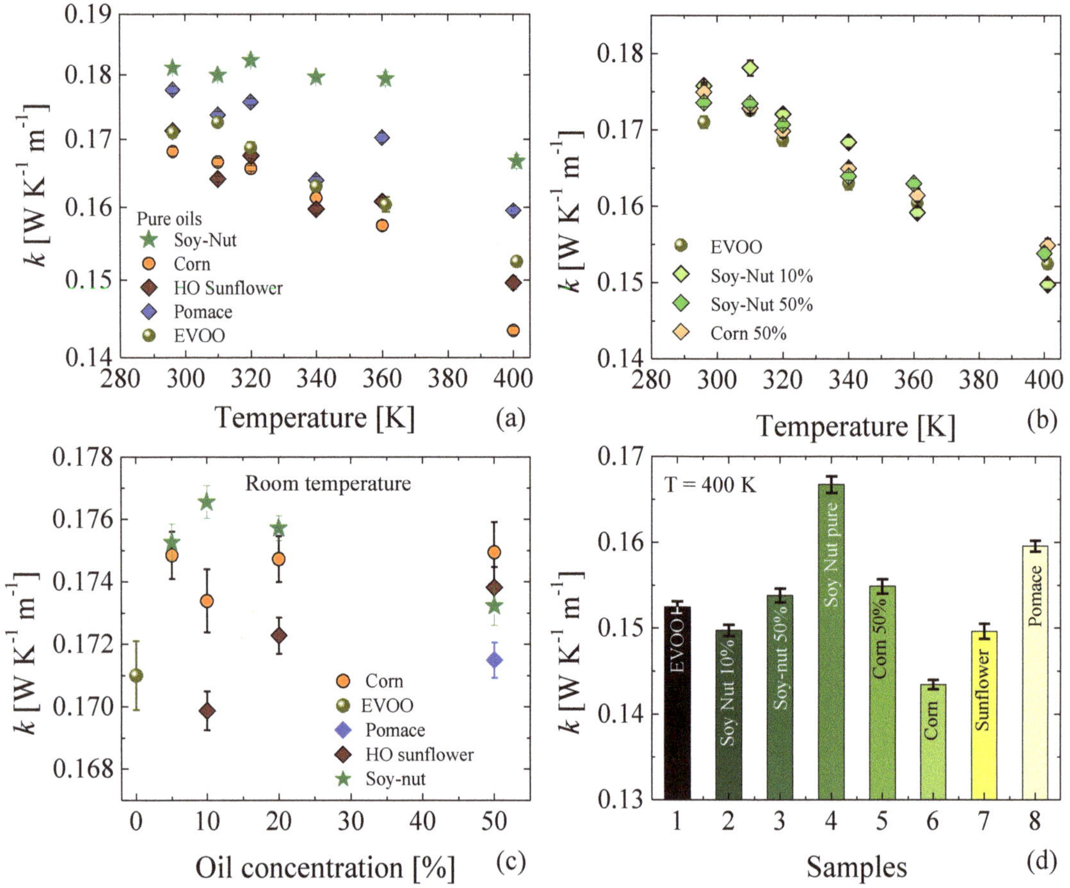

Figure 7. (**a**) Thermal conductivity of pure oils as a function of temperature. (**b**) Thermal conductivity of adulterated oils as a function of temperature. (**c**) Room-temperature thermal conductivity as a function of adulterant oil concentration. (**d**) Comparison of the thermal conductivity at 400 K for different pure oils or adulterated EVOO.

4. Discussion

In this work, the potential of four different characterization techniques to detect traces of adulteration in EVOO were analyzed. Photoluminescence, Raman, and Fourier-transform infrared spectroscopies demonstrated the successful detection of small traces of adulteration on the order of 5%, while the thermal conductivity analysis showed small but constant fluctuations as a function of adulterant oil concentration. Notably, we demonstrated four different characterization methods that are able to rapidly assess the purity of EVOO. Photoluminescence showed a linear decrease in the peak intensity and position as the adulterant oil concentration was increased due to a decrease in the amount of chlorophyll and pheophytin, which are naturally present in EVOO but absent in the adulterant oils. Raman spectroscopy also presented a clear difference between the spectra of EVOO and adulterant oils (even in olive-pomace oil, which is also derived from olives) was also found. Notably, two peaks at ~1155 cm^{-1} and 1523 cm^{-1} were detectable only in EVOO. These modes are associated with the polyene chain of the carotenoids that are naturally presented in EVOO but absent in the adulterants. A clear decrease in the intensity ratio between the peaks at 1523 cm^{-1} (only presented in EVOO) and 1656 cm^{-1} (a common mode presented in all the studied oils) was observed as a function of the adulterant-oil concentration. While a rough comparison between the IR spectra did not show appreciable differences, a statistical analysis showed grouping of the spectra and distinguished a remarkable difference in the PCA scores between pure EVOO and adulterated oils, demonstrating detection of as low as 5% adulterant concentration via FTIR spectroscopy. It is important to note that, while even PCA did not show significant differences between EVOO and EVOO–pomace mixtures, a deeper analysis using a two-dimensional correlation treatment was sensitive to small fluctuations around 2900 cm^{-1}. This result is a nascent effort that demonstrates the potential of 2DCOS analysis for the detection of EVOO adulterated with oils of very similar origin. Finally, an appreciable fluctuation in the thermal conductivity of EVOO was observed for different amounts of adulterant oils. Thermal conductivity has previously been overlooked as a simple but useful figure of merit for assessing food authenticity, but is also a useful manner in which purity can be ascertained. These results highlight the potential of these techniques to detect adulteration, and indicate that the results of the current study can be used as a starting point for the development of spectroscopic methods that allow for the effective and efficient detection of adulteration in olive oils by aiding in identification and classification. While each technique independently may fail to reliably capture small amounts of adulteration in EVOO given the complexity and chemical variability in the oils, a combination of all of them together provides a more holistic base for authentication. For example, as was observed in the case of the FTIR spectra, it is difficult to differentiate EVOO from olive-pomace oils, due to their common origin, though other techniques such as Raman can clearly distinguish the two. Future subsequent development of multiple sensors incorporating and combining these techniques will allow for the acquisition of complete spectral data sets that are critical for precise EVOO authentication. Beyond the authentication of EVOO, the combination of spectroscopic and thermal techniques has the potential to facilitate simplified authentication throughout the food industry.

Supplementary Materials: The following supporting information can be downloaded at: https://www.mdpi.com/article/10.3390/foods11091304/s1, Figure S1: Sample preparation; Figure S2: IR spectra second derivative and pc loading of EVOO, pomace, corn, and soy-nut oils; Figure S3: Photoluminescence of EVOO adulterated with different concentrations of: (a) corn, (b) soy-nut blend, (c) high oleic sunflower, (d) sunflower oils, and (e) olive-pomace oils; Figure S4: Normalized Raman spectra of EVOO adulterated with different concentrations of: (a) ol-ive-pomace, (b) soy-nut blend, and (c) corn oils; Figure S5: Normalized Raman spectra of EVOO measured directly from its package; Figure S6: 2DCOS map of pure EVOO; Table S1: Schematic representation of FTIR dataset for PCA; Table S2: Acid content of the studied edible oils [48].

Author Contributions: Conceptualization, E.C.-A., C.M.S.T. and A.C.-A.; Data curation, E.C.-A., B.P., M.K. and R.C.N.; Formal analysis, E.C.-A., B.P. and M.K.; Funding acquisition, C.M.S.T. and A.C.-A.; Investigation, E.C.-A., B.P. and R.C.N.; Methodology, E.C.-A. and A.C.-A.; Software, E.C.-A. and A.C.-A.; Supervision and validation R.S.F. and C.M.S.T.; Writing—original draft, E.C.-A., R.S.F. and R.C.N.; Writing—review and editing, E.C.-A., R.C.N. and R.S.F. All authors have read and agreed to the published version of the manuscript.

Funding: ICN2 is supported by the Severo Ochoa program from the Spanish Research Agency (AEI, grant no. SEV-2017-0706) and by the CERCA Programme/Generalitat de Catalunya. ICN2 authors acknowledge the support from the Spanish MICINN project SIP (PGC2018-101743-B-I00). A.C.-A. acknowledges the support from Fondecyt Iniciación 11200620. R.C.N. acknowledges funding from the EU-H2020 research and innovation program under the Marie Sklodowska Curie Fellowship (Grant No. 897148).

Institutional Review Board Statement: Not applicable.

Informed Consent Statement: Not applicable.

Data Availability Statement: Raw data can be provided by the corresponding author (ECA) on reasonable request.

Acknowledgments: The infrared-spectroscopy measurements were performed at MIRAS beamline at ALBA Synchrotron with the collaboration of ALBA staff.

Conflicts of Interest: The authors declare no conflict of interest.

References

1. FAO Codex Standars for Fats and Oils from Vegetable Sources. 1999. Available online: http://www.fao.org/docrep/004/y2774e/y2774e04.htm (accessed on 18 October 2021).
2. Dairi, S.; Carbonneau, M.-A.; Galeano-Diaz, T.; Remini, H.; Dahmoune, F.; Aoun, O.; Belbahi, A.; Lauret, C.; Cristol, J.-P.; Madani, K. Antioxidant Effects of Extra Virgin Olive Oil Enriched by Myrtle Phenolic Extracts on Iron-Mediated Lipid Peroxidation under Intestinal Conditions Model. *Food Chem.* **2017**, *237*, 297–304. [CrossRef] [PubMed]
3. Bucciantini, M.; Leri, M.; Nardiello, P.; Casamenti, F.; Stefani, M. Olive Polyphenols: Antioxidant and Anti-Inflammatory Properties. *Antioxidants* **2021**, *10*, 1044. [CrossRef] [PubMed]
4. Fernandes, J.; Fialho, M.; Santos, R.; Peixoto-Plácido, C.; Madeira, T.; Sousa-Santos, N.; Virgolino, A.; Santos, O.; Vaz Carneiro, A. Is Olive Oil Good for You? A Systematic Review and Meta-Analysis on Anti-Inflammatory Benefits from Regular Dietary Intake. *Nutrition* **2020**, *69*, 110559. [CrossRef] [PubMed]
5. Schwingshackl, L.; Lampousi, A.-M.; Portillo, M.P.; Romaguera, D.; Hoffmann, G.; Boeing, H. Olive Oil in the Prevention and Management of Type 2 Diabetes Mellitus: A Systematic Review and Meta-Analysis of Cohort Studies and Intervention Trials. *Nutr. Diabetes* **2017**, *7*, e262. [CrossRef] [PubMed]
6. Guasch-Ferré, M.; Hruby, A.; Salas-Salvadó, J.; Martínez-González, M.A.; Sun, Q.; Willett, W.C.; Hu, F.B. Olive Oil Consumption and Risk of Type 2 Diabetes in US Women. *Am. J. Clin. Nutr.* **2015**, *102*, 479–486. [CrossRef] [PubMed]
7. AlFaris, N.A.; Ba-Jaber, A.S. Effects of a Low-energy Diet with and without Oat Bran and Olive Oil Supplements on Body Mass Index, Blood Pressure, and Serum Lipids in Diabetic Women: A Randomized Controlled Trial. *Food Sci. Nutr.* **2020**, *8*, 3602–3609. [CrossRef]
8. Njike, V.Y.; Ayettey, R.; Treu, J.A.; Doughty, K.N.; Katz, D.L. Post-Prandial Effects of High-Polyphenolic Extra Virgin Olive Oil on Endothelial Function in Adults at Risk for Type 2 Diabetes: A Randomized Controlled Crossover Trial. *Int. J. Cardiol.* **2021**, *330*, 171–176. [CrossRef]
9. Massaro, M.; Scoditti, E.; Carluccio, M.A.; Calabriso, N.; Santarpino, G.; Verri, T.; De Caterina, R. Effects of Olive Oil on Blood Pressure: Epidemiological, Clinical, and Mechanistic Evidence. *Nutrients* **2020**, *12*, 1548. [CrossRef]
10. López-Villodres, J.A.; Abdel-Karim, M.; De La Cruz, J.P.; Rodríguez-Pérez, M.D.; Reyes, J.J.; Guzmán-Moscoso, R.; Rodriguez-Gutierrez, G.; Fernández-Bolaños, J.; González-Correa, J.A. Effects of Hydroxytyrosol on Cardiovascular Biomarkers in Experimental Diabetes Mellitus. *J. Nutr. Biochem.* **2016**, *37*, 94–100. [CrossRef]
11. Sayec, M.L.; Serreli, G.; Diotallevi, C.; Teissier, A.; Deiana, M.; Corona, G. Olive Oil Phenols and Their Metabolites Modulate Nitric Oxide Balance in Human Aortic Endothelial Cells. *Proc. Nutr. Soc.* **2021**, *80*, E34. [CrossRef]
12. Olive World Olive Oil & Health. Available online: https://www.internationaloliveoil.org/olive-world/olive-oil-health/ (accessed on 11 February 2021).
13. Casadei, E.; Valli, E.; Panni, F.; Donarski, J.; Farrús Gubern, J.; Lucci, P.; Conte, L.; Lacoste, F.; Maquet, A.; Brereton, P.; et al. Emerging Trends in Olive Oil Fraud and Possible Countermeasures. *Food Control* **2021**, *124*, 107902. [CrossRef]
14. van Ruth, S.M.; Huisman, W.; Luning, P.A. Food Fraud Vulnerability and Its Key Factors. *Trends Food Sci. Technol.* **2017**, *67*, 70–75. [CrossRef]
15. Conte, L.; Bendini, A.; Valli, E.; Lucci, P.; Moret, S.; Maquet, A.; Lacoste, F.; Brereton, P.; García-González, D.L.; Moreda, W.; et al. Olive Oil Quality and Authenticity: A Review of Current EU Legislation, Standards, Relevant Methods of Analyses, Their Drawbacks and Recommendations for the Future. *Trends Food Sci. Technol.* **2020**, *105*, 483–493. [CrossRef]
16. Bajoub, A.; Bendini, A.; Fernández-Gutiérrez, A.; Carrasco-Pancorbo, A. Olive Oil Authentication: A Comparative Analysis of Regulatory Frameworks with Especial Emphasis on Quality and Authenticity Indices, and Recent Analytical Techniques Developed for Their Assessment. A Review. *Crit. Rev. Food Sci. Nutr.* **2018**, *58*, 832–857. [CrossRef]
17. Tena, N.; Wang, S.C.; Aparicio-Ruiz, R.; García-González, D.L.; Aparicio, R. In-Depth Assessment of Analytical Methods for Olive Oil Purity, Safety, and Quality Characterization. *J. Agric. Food Chem.* **2015**, *63*, 4509–4526. [CrossRef]
18. Kakouri, E.; Revelou, P.-K.; Kanakis, C.; Daferera, D.; Pappas, C.S.; Tarantilis, P.A. Authentication of the Botanical and Geographical Origin and Detection of Adulteration of Olive Oil Using Gas Chromatography, Infrared and Raman Spectroscopy Techniques: A Review. *Foods* **2021**, *10*, 1565. [CrossRef]
19. Barros, I.H.A.S.; Paixão, L.S.; Nascimento, M.H.C.; Lacerda, V.; Filgueiras, P.R.; Romão, W. Use of Portable Raman Spectroscopy in the Quality Control of Extra Virgin Olive Oil and Adulterated Compound Oils. *Vib. Spectrosc.* **2021**, *116*, 103299. [CrossRef]
20. de Lima, T.K.; Musso, M.; Bertoldo Menezes, D. Using Raman Spectroscopy and an Exponential Equation Approach to Detect Adulteration of Olive Oil with Rapeseed and Corn Oil. *Food Chem.* **2020**, *333*, 127454. [CrossRef]
21. Qiu, J.; Hou, H.-Y.; Yang, I.-S.; Chen, X.-B. Raman Spectroscopy Analysis of Free Fatty Acid in Olive Oil. *Appl. Sci.* **2019**, *9*, 4510. [CrossRef]
22. Giovenzana, V.; Beghi, R.; Romaniello, R.; Tamborrino, A.; Guidetti, R.; Leone, A. Use of Visible and near Infrared Spectroscopy with a View to On-Line Evaluation of Oil Content during Olive Processing. *Biosyst. Eng.* **2018**, *172*, 102–109. [CrossRef]

23. Yousef, I.; Ribó, L.; Crisol, A.; Šics, I.; Ellis, G.; Ducic, T.; Kreuzer, M.; Benseny-Cases, N.; Quispe, M.; Dumas, P.; et al. MIRAS: The Infrared Synchrotron Radiation Beamline at ALBA. *Synchrotron Radiat. News* **2017**, *30*, 4–6. [CrossRef]
24. Demsar, J.; Curk, T.; Erjavec, A.; Gorup, C.; Hocevar, T.; Milutinovic, M.; Mozina, M.; Polajnar, M.; Toplak, M.; Staric, A.; et al. Orange: Data Mining Toolbox in Python. *J. Mach. Learn. Res.* **2013**, *14*, 2349–2353.
25. Oh, D.W.; Jain, A.; Eaton, J.K.; Goodson, K.E.; Lee, J.S. Thermal Conductivity Measurement and Sedimentation Detection of Aluminum Oxide Nanofluids by Using the 3ω Method. *Int. J. Heat Fluid Flow* **2008**, *29*, 1456–1461. [CrossRef]
26. Lubner, S.D.; Choi, J.; Wehmeyer, G.; Waag, B.; Mishra, V.; Natesan, H.; Bischof, J.C.; Dames, C. Reusable Bi-Directional 3ω Sensor to Measure Thermal Conductivity of 100-Mm Thick Biological Tissues. *Rev. Sci. Instrum.* **2015**, *86*, 014905. [CrossRef] [PubMed]
27. Jiménez-Pérez, J.L.; Cruz-Orea, A.; Lomelí Mejia, P.; Gutierrez-Fuentes, R. Monitoring the Thermal Parameters of Different Edible Oils by Using Thermal Lens Spectrometry. *Int. J. Thermophys.* **2009**, *30*, 1396–1399. [CrossRef]
28. Lara-Hernández, G.; Suaste-Gómez, E.; Cruz-Orea, A.; Mendoza-Alvarez, J.G.; Sánchez-Sinéncio, F.; Valcárcel, J.P.; García-Quiroz, A. Thermal Characterization of Edible Oils by Using Photopyroelectric Technique. *Int. J. Thermophys.* **2013**, *34*, 962–971. [CrossRef]
29. Gustafsson, S.E.; Karawacki, E.; Khan, M.N. Transient Hot-Strip Method for Simultaneously Measuring Thermal Conductivity and Thermal Diffusivity of Solids and Fluids. *J. Phys. D Appl. Phys.* **1979**, *12*, 1411–1421. [CrossRef]
30. Sergeev, O.A.; Shashkov, A.G.; Umanskii, A.S. Thermophysical Properties of Quartz Glass. *J. Eng. Phys.* **1982**, *43*, 1375–1383. [CrossRef]
31. Turgut, A.; Tavman, I.; Tavman, S. Measurement of Thermal Conductivity of Edible Oils Using Transient Hot Wire Method. *Int. J. Food Prop.* **2009**, *12*, 741–747. [CrossRef]
32. Rojas, E.E.G.; Coimbra, J.S.R.; Telis-Romero, J. Thermophysical Properties of Cotton, Canola, Sunflower and Soybean Oils as a Function of Temperature. *Int. J. Food Prop.* **2013**, *16*, 1620–1629. [CrossRef]
33. Cahill, D.G. Thermal Conductivity Measurement from 30 to 750 K: The 3ω Method. *Rev. Sci. Instrum.* **1990**, *61*, 802. [CrossRef]
34. Cahill, D.G. Erratum: "Thermal Conductivity Measurement from 30 to 750 K: The 3ω Method" [Rev. Sci. Instrum. 61, 802 (1990)]. *Rev. Sci. Instrum.* **2002**, *73*, 3701. [CrossRef]
35. Rodríguez-Laguna, M.R.; Castro-Alvarez, A.; Sledzinska, M.; Maire, J.; Costanzo, F.; Ensing, B.; Pruneda, M.; Ordejón, P.; Sotomayor Torres, C.M.; Gómez-Romero, P.; et al. Mechanisms behind the Enhancement of Thermal Properties of Graphene Nanofluids. *Nanoscale* **2018**, *10*, 15402–15409. [CrossRef]
36. Chavez-Angel, E.; Reuter, N.; Komar, P.; Heinz, S.; Kolb, U.; Kleebe, H.-J.; Jakob, G. Subamorphous Thermal Conductivity of Crystalline Half-Heusler Superlattices. *Nanoscale Microscale Thermophys. Eng.* **2019**, *23*, 1–9. [CrossRef]
37. Lazzerini, C.; Cifelli, M.; Domenici, V. Pigments in Extra-Virgin Olive Oil: Authenticity and Quality. In *Products from Olive Tree*; InTech: London, UK, 2016.
38. Peterson, R.B.; Oja, V.; Laisk, A. Chlorophyll Fluorescence at 680 and 730 Nm and Leaf Photosynthesis. *Photosynth. Res.* **2001**, *70*, 185–196. [CrossRef] [PubMed]
39. Lv, M.Y.; Zhang, X.; Ren, H.R.; Liu, L.; Zhao, Y.M.; Wang, Z.; Wu, Z.L.; Liu, L.M.; Xu, H.J. A Rapid Method to Authenticate Vegetable Oils through Surface-Enhanced Raman Scattering. *Sci. Rep.* **2016**, *6*, 23405. [CrossRef]
40. El-Abassy, R.M.; Donfack, P.; Materny, A. Visible Raman Spectroscopy for the Discrimination of Olive Oils from Different Vegetable Oils and the Detection of Adulteration. *J. Raman Spectrosc.* **2009**, *40*, 1284–1289. [CrossRef]
41. Bernstein, P.S. New Insights into the Role of the Macular Carotenoids in Age-Related Macular Degeneration. Resonance Raman Studies. *Pure Appl. Chem.* **2002**, *74*, 1419–1425. [CrossRef]
42. Withnall, R.; Chowdhry, B.Z.; Silver, J.; Edwards, H.G.M.; de Oliveira, L.F.C. Raman Spectra of Carotenoids in Natural Products. *Spectrochim. Acta Part A Mol. Biomol. Spectrosc.* **2003**, *59*, 2207–2212. [CrossRef]
43. Zeb, A.; Murkovic, M. Carotenoids and Triacylglycerols Interactions during Thermal Oxidation of Refined Olive Oil. *Food Chem.* **2011**, *127*, 1584–1593. [CrossRef]
44. Concha-Herrera, V.; Lerma-GarcÍa, M.J.; Herrero-MartiÍnez, J.M.; SimoÓ-Alfonso, E.F. Prediction of the Genetic Variety of Extra Virgin Olive Oils Produced at La Comunitat Valenciana, Spain, by Fourier Transform Infrared Spectroscopy. *J. Agric. Food Chem.* **2009**, *57*, 9985–9989. [CrossRef] [PubMed]
45. Vanstone, N.; Moore, A.; Martos, P.; Neethirajan, S. Detection of the Adulteration of Extra Virgin Olive Oil by Near-Infrared Spectroscopy and Chemometric Techniques. *Food Qual. Saf.* **2018**, *2*, 189–198. [CrossRef]
46. Qiu, J.; Hou, H.-Y.; Huyen, N.T.; Yang, I.-S.; Chen, X.-B. Raman Spectroscopy and 2DCOS Analysis of Unsaturated Fatty Acid in Edible Vegetable Oils. *Appl. Sci.* **2019**, *9*, 2807. [CrossRef]
47. Lasch, P.; Noda, I. Two-Dimensional Correlation Spectroscopy (2D-COS) for Analysis of Spatially Resolved Vibrational Spectra. *Appl. Spectrosc.* **2019**, *73*, 359–379. [CrossRef] [PubMed]
48. FAO. *Codex Standard for Fats and Oils from Vegetable Sources*; FAO: Rome, Italy, 1999.

Article

Predicting Satiety from the Analysis of Human Saliva Using Mid-Infrared Spectroscopy Combined with Chemometrics

Dongdong Ni, Heather E. Smyth, Michael J. Gidley and Daniel Cozzolino *

Centre for Nutrition and Food Sciences, Queensland Alliance for Agriculture and Food Innovation, The University of Queensland, St Lucia, QLD 4072, Australia; d.ni@uq.edu.au (D.N.); h.smyth@uq.edu.au (H.E.S.); m.gidley@uq.edu.au (M.J.G.)
* Correspondence: d.cozzolino@uq.edu.au

Abstract: The aim of this study was to evaluate the ability of mid-infrared (MIR) spectroscopy combined with chemometrics to analyze unstimulated saliva as a method to predict satiety in healthy participants. This study also evaluated features in saliva that were related to individual perceptions of human–food interactions. The coefficient of determination (R^2) and standard error in cross validation (SECV) for the prediction of satiety in all saliva samples were 0.62 and 225.7 satiety area under the curve (AUC), respectively. A correlation between saliva and satiety was found, however, the quantitative prediction of satiety using unstimulated saliva was not robust. Differences in the MIR spectra of saliva between low and high satiety groups, were observed in the following frequency ratios: 1542/2060 cm^{-1} (total protein), 1637/3097 cm^{-1} (α-amino acids), and 1637/616 (chlorides) cm^{-1}. In addition, good to excellent models were obtained for the prediction of satiety groups defined as low or high satiety participants (R^2 0.92 and SECV 0.10), demonstrating that this method could be used to identify low or high satiety perception types and to select participants for appetite studies. Although quantitative PLS calibration models were not achieved, a qualitative model for the prediction of low and high satiety perception types was obtained using PLS-DA. Furthermore, this study showed that it might be possible to evaluate human/food interactions using MIR spectroscopy as a rapid and cost-effective tool.

Keywords: saliva; spectroscopy; satiety; satiation; chemometrics

Citation: Ni, D.; Smyth, H.E.; Gidley, M.J.; Cozzolino, D. Predicting Satiety from the Analysis of Human Saliva Using Mid-Infrared Spectroscopy Combined with Chemometrics. *Foods* **2022**, *11*, 711. https://doi.org/10.3390/foods11050711

Academic Editor: Adriana Franca

Received: 10 February 2022
Accepted: 24 February 2022
Published: 28 February 2022

1. Introduction

A wide range of instrumental and spectroscopic methods such as nuclear magnetic resonance (NMR) [1,2], mass spectrometry (MS) [3], and vibrational spectroscopy (e.g., mid-infrared, Raman) [4,5], have been utilized to profile and analyze the biochemical and chemical composition of saliva [6,7]. Vibrational spectroscopy, and in particular infrared (IR) spectroscopy, has attracted the attention of researchers due to its high sensitivity, high speed, and low cost. The utilization of vibrational spectroscopy has been reported in many areas, including medicine, chemistry, forensic, and food sciences, for the measurement of composition and functional properties [8–10]. During the last 20 years, advances in instrumentation and computing have allowed for the evolution of diagnostic methodologies based in vibrational spectroscopy including near infrared (NIR), mid-infrared (MIR), and Raman spectroscopy of saliva and other biofluids [11–14].

It is well known that saliva relates to oral physiology and plays an important role in both oral processing and sensory perception of food [15–18]. Several studies have reported that salivary proteins are not only correlated with the fungiform papillae density of tongues [19], but they also associated with food aroma compounds released during food intake. Salivary proteins also have a role in the mediation of taste components in sensory perception [17,18]. These factors have strong effects on oral food processing, influencing food particle size, food-saliva interactions, and how nutrients are released during the process of food intake [20]. Moreover, recent studies were focused on the effects of these factors (e.g., oral food processing, food particle size, and food sensory perceptions) on food intake, satiation, and satiety [15,21,22]. A recent study [15] evaluated the relationships between oral sensory exposure and hormones involved in longer oral processing and how these relationships might have influenced satiation and food intake in humans [15,21,22]. It is well known that saliva contains information not only about the composition of the food, but also on the biological and physical properties where recently studies [23,24] reported saliva might be associated with body composition and energy expenditure, which were highly associated with satiety.

Satiation and satiety are defined as perceived fullness feelings during and after a meal and they are considered the main driving forces responsible for the control of eating behaviors in humans [25–27]. The roles of satiation and satiety are to modulate daily meal portion sizes and frequency. Satiety has been proposed to be a key factor in controlling food intake with impacts on an individual's ability to manage their nutrition and body condition (e.g., weight) [25–27]. Moreover, satiation and satiety are not only related to food and energy intake, but they also influence psychological status (e.g., emotion and mood) [28,29]. For example, uncontrolled hunger and psychological phenomena, such as feelings of deprivation and cravings, are major difficulties and main reasons for the ultimate failure of keeping a healthy diet [28,29]. Low satiety human phenotypes [30,31] were also reported in the literature and defined as individuals who can recognize appetite sensations, with both low appetite and low changes in appetite sensation. This research suggested that low satiety individuals have lower self-reported satiety than the high satiety individuals. Drapeau and colleagues [30] reported that low satiety phenotype individuals had a lower blood cortisol response during meal intake. However, only limited physiological differences were reported for this phenotype, as it was challenging to identify this phenotype before the experiment.

The aim of this study was to assess the ability of mid-infrared spectroscopy to predict satiety (and satiation) using the saliva collected from healthy participants. This approach will open the possibility of utilizing rapid and low-cost methods, such as IR spectroscopy, to evaluate human/food interactions (satiation and satiety) as well as to explore the technology as a screening method in other food and/or physiological studies.

2. Materials and Methods

2.1. Participants and Saliva Collection

A total of 52 healthy participants (31 female and male 21, with an average of 38.1 and standard deviation of ±13.8 years) were recruited through an open advertisement posted around Brisbane city (Brisbane, QLD, Australia). The selection criteria for participants were based on the following parameters: aged between 18 and 70 years, lack of oral cavities or dental diseases, no diabetes, not being pregnant or lactating at the time of the experiments, and not being diagnosed with psychological diseases, such as depression. The ethics for the research was approved by the Sub-Committee for Human Research Ethics of the University of Queensland (approval number: 2019002688).

Unstimulated saliva, defined as the saliva that continuously bathes the oral cavity without chewing and taste stimulation [32,33], was collected from each participant prior to the satiation and satiety sensory experiment three times during three consecutive days. On the experimental day, no food or drink (except water) were allowed after breakfast (the recommended consumption time for breakfast was 7:00 a.m. for all the participants). Participants were asked to arrive at the sensory lab around 10:00 a.m. after 3 h fasting. Details about the protocols and methods used to collect the unstimulated saliva were reported in a previous paper [34]. In short, a saliva collection suite was provided to each participant, which included a cup of 10 mL mouth rinsing water, a saliva collection tube stored in a beaker with filled with ice, and a spittoon. Participants were asked to rinse their mouths properly with the provided water, spit the water into the spittoon, and avoid swallowing the saliva for 2 min where they could expectorate saliva into the collection tube every 30 s. The expectorated saliva samples were sealed, double wrapped with plastic bags, and transferred into the −80 °C freezer. In total 156 saliva samples were collected (52 participants × 3 days).

2.2. Satiation and Satiety Measurements

Three types of plant-based foods were utilized during the satiation and satiety sensory experiments: an apple (Royal Gala variety); a banana (Cavendish variety); and an avocado (Hass variety). The selection of plant-based foods (e.g., vegetables, fruits) used in this study was related to the fact that plant foods take up over two thirds of the everyday diet recommendations worldwide. The key features of these plant-based foods included high dietary fiber and cell wall structuring that tends to induce both strong satiation and long-lasting satiety. The experiments were conducted on three different days (one food type each day). The three plant-based foods were selected as they each provide a different source of energy, such as soluble sugars for the apple, starch for the banana, and lipids for the avocado). These differences might have induced different satiation and satiety in the participants. The experimental design followed in this study not only balanced the variations in nutrients and energy sources, but also allowed for the comparison of different food types. Each food was cut into approximately 3 mm slices and served in a covered plastic container (100 g portions in each container). Food was served ad libitum during 20 min for each participant. Participants could stop eating when they felt comfortably full. The fullness ratings were determined as described in a previously published paper [35]. Participants' fullness during the meal and for 150 min after the meal were self-evaluated using a 20 cm labelled magnitude scale (LMS) [36]. Fullness ratings at the time points were measured by the distance (cm) between the greatest imaginable hunger point and the participant's marked point on the scale. The total area under the curve (AUC) of the perceived fullness over time was defined as satiation (AUC during the meal) and satiety (AUC after the meal) [27]. The participants were separated into low or high satiety groups according to the satiety values (AUC). The low satiety perceiver group were defined as those having an AUC lower than 1100 AUC, whereas the high satiety perceiver group contained the participants with an AUC that was higher than 1100.

2.3. Mid-Infrared Spectrum Collection for Saliva

The spectrum of the saliva samples was collected using an MIR spectrometer ALPHA II (Bruker Optics, Ettilgen, Germany) (4000 cm^{-1} to 400 cm^{-1} region). The spectrometer was equipped with a diamond-attenuated total reflection (ATR) crystal. Frozen saliva samples were thawed at room temperature (25 °C) for half an hour and the thawed saliva samples were homogenized using a vortex (2000 rpm for 20 s) prior to the MIR measurement. The ATR crystal was fully covered with the homogenized saliva sample (approx. 5 μL). Samples were immediately scanned, and spectra recorded. Each spectrum was computed using an average of 24 co-added interferograms at a resolution of 4 cm^{-1}. A spectrum of air was used as a background prior to sample measurement and the spectrum of water was also measured every 20 samples. The instrument was operated using the OPUS software (version 8.5.29, Bruker Optics, Ettilgen, Germany). After each measurement, the surface of the ATR crystal was cleaned utilizing a 70% *w*/*w* ethanol/water solution and wiped with tissue paper between samples. A total of 156 saliva samples were collected for further analysis.

2.4. Chemometric Analysis

Before chemometric analysis, the MIR data of the unstimulated saliva samples were pre-processed using a baseline correction and Savitzky–Golay second derivative (second polynomial order and 21 smoothing points) (The Unscrambler X, Camo, Oslo, Norway) [37,38]. The fingerprint region (1800 to 450 cm^{-1}) was utilized to establish partial least squares regression (PLS) models to predict satiety and satiation in the saliva samples using the MIR spectra. Classification models for low and high satiety were also developed using PLS discriminant analysis (DA) where samples belonging to the low satiety group were identified with the number 1 and samples from the high satiety group were identified with the number 2. The threshold governed the choice to turn a projected probability or score into a class label. In this study, the threshold was set to 1.5. The PLS and PLS-DA models developed were validated using cross validation (leave-one-out) [39,40].

3. Results and Discussion

Figure 1A shows the average MIR spectrum of the collected saliva samples in the fingerprint region (1800 to 600 cm^{-1}) and compares the low and high satiety groups, as well as the saliva collected from all participants where all the food types analyzed were included. The effect of food types on the high or low satiety responses were observed in the MIR spectra of the saliva collected. In both avocado and banana, the absorbances for the low satiety perceiver samples were lower than the high satiety perceivers in the fingerprint region (1800–600 cm^{-1}). The low absorbances might have indicated that the low satiety perceivers generally tended to produce more diluted saliva than the participants in the higher perceivers group. The main differences in the absorption values were observed between 1336–1364 cm^{-1} and were associated with the stretching vibrations of the carboxyl groups COO and asymmetric C-N stretching, which corresponded with the amide III group [5,41–44]. At 1270 cm^{-1} this band could be associated with CO groups corresponding with the presence of esters, and around 1076 cm^{-1} was associated with the presence of glycosylated proteins and phosphorus-containing compounds [5,41,42,44]. Peaks around 1437 cm^{-1} and 1473 cm^{-1} were associated with vibrations of $\delta(CH_2)$ groups corresponding with proteins, lipids, fatty acids, and polysaccharides. These peaks have also been reported as biochemical indicators for triene conjugates and superoxide dismutase, which are present in saliva [32]. The peak at 1542 cm^{-1} has been reported to be associated with amide II (δNH, νCN) groups corresponding with salivary seromucoids [32]. The peak at 1647 cm^{-1} could be associated with amide I corresponding with albumin in the saliva, whereas the peak at 1653 cm^{-1} has been reported to be associated with amide I proteins in an α-helix conformation for salivary proteins [32]. The peak at 1717 cm^{-1} was associated with amide I purine bases, DNA, and RNA [5,32,41,42,44].

It has been reported [32] that the ratios between specific absorption bands in the spectra of saliva are correlated with salivary biochemical indicators (e.g., total protein, α-amino acids, lactate dehydrogenase). Figure 1B–D shows the ratios between frequencies at specific peaks when comparing high and low satiety perceivers for each of the three food types analyzed. When sorting out the saliva-satiety data according to the specific food types, differences could be identified. The low satiety perceivers had higher values for ratios 1542/2060 cm^{-1}, 1637/3097 cm^{-1}, and 1637/1616 cm^{-1} for avocado ($p < 0.05$) than the high satiety perceivers. Although differences in the same direction were found for apple and banana, these were not statistically significantly different. Similar results were reported by other authors where the ratio between 1542/2060 cm^{-1} was associated with total proteins based on the amide II group band and SCN^- thiocyanate in the saliva, between 1637/3097 cm^{-1} and 1637/1616 cm^{-1}, was associated with α-amino acids and chlorides, respectively [32]. Although saliva from low satiety perceivers was more diluted (more water), it had an apparently higher percentage of protein and amino acids compared with other salivary organic components. The observed differences in the spectra of the saliva might have indicated that compositional variations in human saliva may be a result of underlying factors related to satiety perception types. One explanation could be that the saliva of low satiety individuals was more watery or diluted, but the concentration of proteins, α-amino acids, or chlorides, was higher compared to the high satiety saliva samples [33].

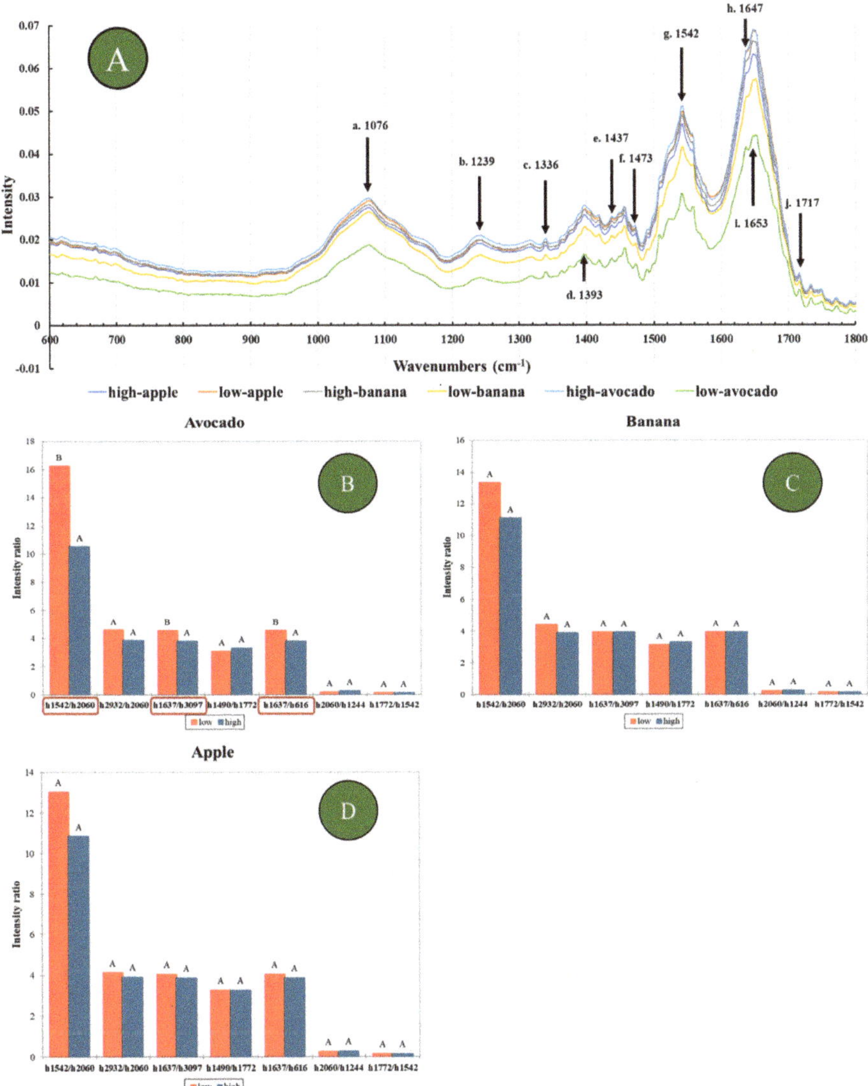

Figure 1. Mid-infrared spectra and ratios at specific frequencies of the unstimulated saliva samples analysed to show differences between low and high satiety groups. (**A**) Fingerprint region of salivary spectra comparing both food types and satiety perception types. The main reported absorption peaks in the literature [5,32,33,41–46] were labelled with lower case and wavelength number (a1076 cm^{-1}, glycosylated proteins and phosphorus-containing components; b1239 cm^{-1}, amide III/phospholipids; c1336 cm^{-1}, carboxyl groups COO and asymmetric C-N stretching; d1393 cm^{-1}, asymmetric and symmetric CH2 bending; e1437 cm^{-1}, and f1473 cm^{-1}, $\delta(CH_2)$ groups corresponding to biochemical indicators for triene conjugates and superoxide dismutase; g1542 cm^{-1}, amide II (δNH, νCN) groups; h1647 cm^{-1}, amide I corresponding with albumin; i1653 cm^{-1}, amide I proteins in α-helix; and j1717 cm^{-1}, amide I purine bases, DNA and RNA). (**B**) Avocado, (**C**) banana, and (**D**) apple; ratios at specific frequencies calculated from the salivary spectra comparing the high and low satiety perceiver groups. The capital letters (e.g., A and B) in the figures signify significant difference between satiety perception groups.

Table 1 shows the PLS cross validation calibration statistics for the prediction of satiety and satiation using unstimulated saliva collected from healthy participants. The coefficient of determination in cross validation (R^2) and standard error in cross validation (SECV) reported for the prediction of satiety in all samples was 0.62 and 225.7 AUC, respectively. In contrast, poor calibration models were obtained for the prediction of satiation ($R^2 < 0.20$) in all samples. These results might have reflected the nature of the experiment in which participants were asked to eat until comfortably full; although, there were differences in satiation responses between individuals where the term 'comfortably' full was interpreted differently by individuals. The lack of correlation with the MIR spectra suggested that the different responses were unlikely to be due to differences in oral physiology, as was reflected in the saliva composition. Calibration models were also developed for the different food types used. The R^2 and SECV obtained for the prediction of satiety after consuming a banana was 0.63 and 188.1 AUC, respectively. However, poor PLS calibration models were obtained for the prediction of satiety using saliva samples collected from either the apple or avocado experiments ($R^2 < 0.20$).

Table 1. Descriptive statistics, partial least squares regression cross validation statistics for the prediction of satiety in saliva samples, and the PLS-DA cross validation statistics for the classification of saliva as low or high satiety.

	All Foods	Banana	Avocado	PLS-DA
R^2	0.62	0.63	0.20	0.92
SECV	225.7	188.1	237.5	0.10
Bias	4.72	−12.5	0.60	0.001
Slope	0.67	0.62	0.20	0.97
LV	7	8	1	11
Mean (AUC)	1363	1456	1368	
SD	409	472	319	
Range	3138–423	3138–707	2272–525	

PLS-DA: partial least squares discriminant analysis; R^2: coefficient of determination in calibration (R^2); SECV: standard error in cross validation; SD: standard deviation; LV: number of latent variables used to develop the models.

The highest PLS loadings (Figure 2) for the prediction of satiety using all samples were observed around 1750 cm^{-1}, which was associated with phospholipid, lipid, and ester groups [45,46], between 1665–1616 cm^{-1} was associated with amide I groups (proteins), 1286 cm^{-1} was associated with amide III groups, 1247 cm^{-1} was associated with PO_2 of phosphate, 1089 cm^{-1} was associated with the symmetric stretching of phosphate groups of phosphodiester, 989 cm^{-1} was associated with C-O of ribose and C-C bonds, and 957 cm^{-1} was associated with polysaccharides [5,41–44]. Ni et al. [16] reported that the MIR frequencies between 1766–1725 cm^{-1} and 1692–1632 cm^{-1} were the most important when MIR calibrations were developed for the prediction of saliva flow, oral processing time, and fungiform papillae density of tongue. Other authors have reported that these oral physiology variables contributed to explaining satiety [15,22,47].

Previous studies have also reported on the prevalence of the low satiety phenotype groups in humans [30,31]. These researchers indicated that the presence of this phenotype group could be associated with stress and anxiety or lower blood cortisol responses to the meal. As described in Section 2, in this study, two groups were defined based on the satiety values (low and high). The R^2 and SECV obtained for the prediction of the low and high satiety group was 0.92 and 0.10, respectively. The PLS-DA results showed that 100% and 98% of the saliva samples were correctly classified as low and high satiety perceivers, respectively.

Overall, this study showed that the use of MIR spectroscopy provided a practical tool to understand the complex relationships between human physiology and self-reported responses based on human-food interactions. The results also showed that a relationship between saliva composition and satiety existed, although the quantitative models for the prediction of satiety were not robust. However, the use of PLS-DA models allowed reliable identification of saliva samples sourced from participants having low or high satiety responses. These models also indicated that MIR spectroscopy could be used for pre-selection or screening of participants in appetite sensory studies, reducing the time and cost of these types of studies.

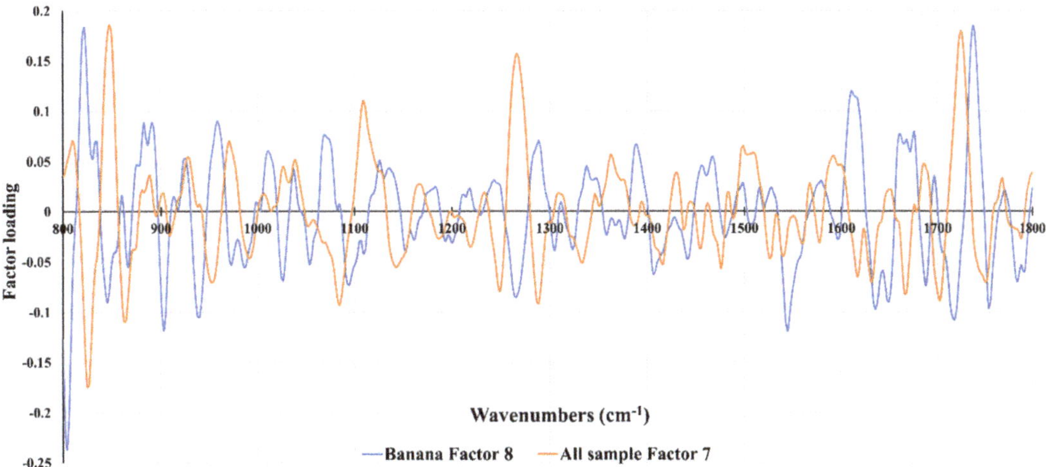

Figure 2. Partial least squares loadings derived from the calibration models used to predict satiety in the saliva of all samples or in the banana samples.

Some of the underlying factors that might influence the performance of the calibration models could be attributed to the fact that saliva itself was only one of many potential factors that could be considered to evaluate the human status to predict appetite. The human experience of appetite is not only decided by the human condition but is also influenced by the environment and whether the experiment design and saliva collection protocols were adequate to evaluate human–food interactions using MIR spectroscopy. Another important factor to consider was the utilization of unstimulated saliva. In this study, unstimulated saliva was used; however, it was possible that the use of stimulated saliva (e.g., after exposure to a specific chemical or mechanical stimulus) would result in alternate, or possibly more targeted, information on food-satiety interactions using MIR spectroscopy.

4. Conclusions

Results from this study demonstrated the ability of MIR spectroscopy combined with chemometrics (e.g., PLS) to predict satiety from resting (unstimulated) saliva samples. Although quantitative PLS calibration models were not achieved, a qualitative model for the prediction of low and high satiety perception type was obtained using PLS-DA. Furthermore, this study indicated the possibility of evaluating the interactions between saliva and food using MIR spectroscopy as a rapid and cost-effective tool.

Author Contributions: D.N., formal analysis, writing—original draft preparation, writing—review and editing; H.E.S., supervision, writing—review and editing; M.J.G., supervision, writing—review and editing; D.C., methodology, formal analysis, writing—original draft preparation, writing—review and editing. All authors have read and agreed to the published version of the manuscript.

Funding: Scholarship from the China Scholarship Council and the University of Queensland.

Institutional Review Board Statement: Not applicable.

Informed Consent Statement: Not applicable.

Data Availability Statement: Not applicable.

Acknowledgments: The authors acknowledge funding from Hort Innovation (Australia). Dongdong Ni acknowledges the award of a scholarship from the China Scholarship Council and the University of Queensland. The sensory panel from the Health and Food Sciences Precinct (Coopers Plains, Queensland, Australia) are acknowledged for their dedication and participation in this research.

Conflicts of Interest: The authors declare no conflict of interest.

References

1. Gardner, A.; Parkes, H.G.; Carpenter, G.H.; So, P.W. Developing and Standardizing a Protocol for Quantitative Proton Nuclear Magnetic Resonance (^{1}H NMR) Spectroscopy of Saliva. *J. Proteome Res.* **2018**, *17*, 1521–1531. [CrossRef]
2. Figueira, J.; Jonsson, P.; Adolfsson, A.N.; Adolfsson, R.; Nyberg, L.; Ohman, A. NMR analysis of the human saliva metabolome distinguishes dementia patients from matched controls. *Mol. BioSyst.* **2016**, *12*, 2562–2571. [CrossRef] [PubMed]
3. De Filippis, F.; Vannini, L.; La Storia, A.; Laghi, L.; Piombino, P.; Stellato, G.; Serrazanetti, D.I.; Gozzi, G.; Turroni, S.; Ferrocino, I.; et al. The same microbiota and a potentially discriminant metabolome in the saliva of omnivore, ovo-lacto-vegetarian and Vegan individuals. *PLoS ONE* **2014**, *9*, e112373. [CrossRef] [PubMed]
4. Muro, C.K.; Fernandes, L.D.; Lednev, I.K. Sex Determination Based on Raman Spectroscopy of Saliva Traces for Forensic Purposes. *Anal. Chem.* **2016**, *88*, 12489–12493. [CrossRef]
5. Talari, A.C.S.; Martinez, M.A.G.; Movasaghi, Z.; Rehman, S.; Rehman, I.U. Advances in Fourier transform infrared (FTIR) spectroscopy of biological tissues. *Appl. Spectrosc. Rev.* **2017**, *52*, 456–506. [CrossRef]
6. Pereira, J.L.; Duarte, D.; Carneiro, T.J.; Ferreira, S.; Cunha, B.; Soares, D.; Costa, A.L.; Gil, A.M. Saliva NMR metabolomics: Analytical issues in pediatric oral health research. *Oral Dis.* **2019**, *25*, 1545–1554. [CrossRef]
7. Mikkonen, J.J.; Raittila, J.; Rieppo, L.; Lappalainen, R.; Kullaa, A.M.; Myllymaa, S. Fourier Transform Infrared Spectroscopy and Photoacoustic Spectroscopy for Saliva Analysis. *Appl. Spectrosc.* **2016**, *70*, 1502–1510. [CrossRef] [PubMed]
8. Orphanou, C.M.; Walton-Williams, L.; Mountain, H.; Cassella, J. The detection and discrimination of human body fluids using ATR FT-IR spectroscopy. *Forensic Sci. Int.* **2015**, *252*, e10–e16. [CrossRef] [PubMed]
9. Graca, G.; Moreira, A.S.; Correia, A.J.V.; Goodfellow, B.J.; Barros, A.S.; Duarte, I.F.; Carreira, I.M.; Galhano, E.; Pita, C.; Almeida, M.D.; et al. Mid-infrared (MIR) metabolic fingerprinting of amniotic fluid: A possible avenue for early diagnosis of prenatal disorders? *Anal. Chim. Acta* **2013**, *764*, 24–31. [CrossRef]
10. Khaustova, S.; Shkurnikov, M.; Tonevitsky, E.; Artyushenko, V.; Tonevitsky, A. Noninvasive biochemical monitoring of physiological stress by Fourier transform infrared saliva spectroscopy. *Analyst* **2010**, *135*, 3183–3192. [CrossRef]
11. Bec, K.B.; Grabska, J.; Huck, C.W. Near-Infrared Spectroscopy in Bio-Applications. *Molecules* **2020**, *25*, 2948. [CrossRef] [PubMed]
12. Zlotogorski-Hurvitz, A.; Dekel, B.; Malonek, D.; Yahalom, R.; Vered, M. FTIR-based spectrum of salivary exosomes coupled with computational-aided discriminating analysis in the diagnosis of oral cancer. *J. Cancer Res. Clin. Oncol.* **2019**, *145*, 685–694. [CrossRef] [PubMed]
13. Scott, D.A.; Renaud, D.E.; Krishnasamy, S.; Meric, P.; Buduneli, N.; Cetinkalp, S.; Liu, K.Z. Diabetes-related molecular signatures in infrared spectra of human saliva. *Diabetol. Metab. Syndr.* **2010**, *2*, 1–9. [CrossRef] [PubMed]
14. Wongkamhaeng, K.; Poachanukoon, O.; Koontongkaew, S. Dental caries, cariogenic microorganisms and salivary properties of allergic rhinitis children. *Int. J. Pediatr. Otorhinolaryngol.* **2014**, *78*, 860–865. [CrossRef]
15. Lasschuijt, M.; Mars, M.; de Graaf, C.; Smeets, P.A.M. How oro-sensory exposure and eating rate affect satiation and associated endocrine responses-a randomized trial. *Am. J. Clin. Nutr.* **2020**, *111*, 1137–1149. [CrossRef]
16. Ni, D.; Smyth, H.E.; Gidley, M.J.; Cozzolino, D. Exploring the relationships between oral sensory physiology and oral processing with mid infrared spectra of saliva. *Food Hydrocoll.* **2021**, *120*, 106896. [CrossRef]
17. Ployon, S.; Brule, M.; Andriot, I.; Morzel, M.; Canon, F. Understanding retention and metabolization of aroma compounds using an in vitro model of oral mucosa. *Food Chem.* **2020**, *318*, 126468. [CrossRef]
18. Canon, F.; Neiers, F.; Guichard, E. Saliva and Flavor Perception: Perspectives. *J. Agric. Food Chem.* **2018**, *66*, 7873–7879. [CrossRef]
19. Gardner, A.; Carpenter, G.H. Anatomical stability of human fungiform papillae and relationship with oral perception measured by salivary response and intensity rating. *Sci. Rep.* **2019**, *9*, 9759. [CrossRef]
20. Mosca, A.C.; Chen, J.S. Food-saliva interactions: Mechanisms and implications. *Trends Food Sci. Technol.* **2017**, *66*, 125–134. [CrossRef]
21. Zijlstra, N.; de Wijk, R.A.; Mars, M.; Stafleu, A.; de Graaf, C. Effect of bite size and oral processing time of a semisolid food on satiation. *Am. J. Clin. Nutr.* **2009**, *90*, 269–275. [CrossRef] [PubMed]
22. Hogenkamp, P.S.; Schiöth, H.B. Effect of oral processing behaviour on food intake and satiety. *Trends Food Sci. Technol.* **2013**, *34*, 67–75. [CrossRef]
23. Goloni, C.; Peres, F.M.; Senhorello, I.L.S.; Di Santo, L.G.; Mendonca, F.S.; Loureiro, B.A.; Pfrimer, K.; Ferriolli, E.; Pereira, G.T.; Carciofi, A.C. Validation of saliva and urine use and sampling time on the doubly labelled water method to measure energy expenditure, body composition and water turnover in male and female cats. *Br. J. Nutr.* **2020**, *124*, 457–469. [CrossRef]

24. Pruszkowska-Przybylska, P.; Sitek, A.; Rosset, I.; Sobalska-Kwapis, M.; Slomka, M.; Strapagiel, D.; Zadzinska, E.; Morling, N. Association of saliva 25(OH)D concentration with body composition and proportion among pre-pubertal and pubertal Polish children. *Am. J. Hum. Biol.* **2020**, *32*, e23397. [CrossRef] [PubMed]
25. De Graaf, C.; Blom, W.A.M.; Smeets, P.A.M.; Stafleu, A.; Hendriks, H.F.J. Biomarkers of satiation and satiety. *Am. J. Clin. Nutr.* **2004**, *79*, 946–961. [CrossRef] [PubMed]
26. Gibbons, C.; Hopkins, M.; Beaulieu, K.; Oustric, P.; Blundell, J.E. Issues in Measuring and Interpreting Human Appetite (Satiety/Satiation) and Its Contribution to Obesity. *Curr. Obes. Rep.* **2019**, *8*, 77–87. [CrossRef] [PubMed]
27. Blundell, J.; de Graaf, C.; Hulshof, T.; Jebb, S.; Livingstone, B.; Lluch, A.; Mela, D.; Salah, S.; Schuring, E.; van der Knaap, H.; et al. Appetite control: Methodological aspects of the evaluation of foods. *Obes. Rev.* **2010**, *11*, 251–270. [CrossRef]
28. Higgs, S.; Spetter, M.S. Cognitive Control of Eating: The Role of Memory in Appetite and Weight Gain. *Curr. Obes. Rep.* **2018**, *7*, 50–59. [CrossRef]
29. Beaulieu, K.; Blundell, J. The Psychobiology of Hunger—A Scientific Perspective. *Topoi* **2020**, *40*, 565–574. [CrossRef]
30. Drapeau, V.; Blundell, J.; Gallant, A.R.; Arguin, H.; Despres, J.P.; Lamarche, B.; Tremblay, A. Behavioural and metabolic characterisation of the low satiety phenotype. *Appetite* **2013**, *70*, 67–72. [CrossRef]
31. Drapeau, V.; Hetherington, M.; Tremblay, A. Impact of eating and lifestyle behaviors on body weight: Beyond energy value. In *Handbook of Behavior, Food and Nutrition*; Springer: Berlin/Heidelberg, Germany, 2011; pp. 693–706.
32. Bel'skaya, L.V.; Sarf, E.A. Biochemical composition and characteristics of salivary FTIR spectra: Correlation analysis. *J. Mol. Liq.* **2021**, *341*, 117380. [CrossRef]
33. Stading, M.; Johansson, D.; Diogo Löfgren, C.; Christersson, C. Viscoelastic properties of saliva from different glands. In Proceedings of the Nordic Rheology Conference (NRC), Reykjavík, Iceland, 19–21 August 2009; pp. 109–112.
34. Stokes, J.R.; Davies, G.A. Viscoelasticity of human whole saliva collected after acid and mechanical stimulation. *Biorheology* **2007**, *44*, 141–160. [PubMed]
35. Ni, D.; Gunness, P.; Smyth, H.E.; Gidley, M.J. Exploring relationships between satiation, perceived satiety, and plant-based snack food features. *Int. J. Food Sci. Technol.* **2021**, *56*, 5340–5351. [CrossRef]
36. Zalifah, M.K.; Greenway, D.R.; Caffin, N.A.; D'Arcy, B.R.; Gidley, M.J. Application of labelled magnitude satiety scale in a linguistically-diverse population. *Food Qual. Prefer.* **2008**, *19*, 574–578. [CrossRef]
37. Movasaghi, Z.; Rehman, S.; Rehman, I.U. Fourier transform infrared (FTIR) spectroscopy of biological tissues. *Appl. Spectrosc. Rev.* **2008**, *43*, 134–179. [CrossRef]
38. Savitzky, A.; Golay, M.J. Smoothing and differentiation of data by simplified least squares procedures. *Anal. Chem.* **1964**, *36*, 1627–1639. [CrossRef]
39. Næs, T.; Isaksson, T.; Fearn, T.; Davies, T. *A User-Friendly Guide to Multivariate Calibration and Classification*, 2nd ed.; NIR: Chichester, UK, 2002; Volume 6.
40. Bureau, S.; Cozzolino, D.; Clark, C.J. Contributions of Fourier-transform mid infrared (FT-MIR) spectroscopy to the study of fruit and vegetables: A review. *Postharvest Biol. Technol.* **2019**, *148*, 1–14. [CrossRef]
41. Rodrigues, R.P.; Aguiar, E.M.; Cardoso-Sousa, L.; Caixeta, D.C.; Guedes, C.C.; Siqueira, W.L.; Maia, Y.C.P.; Cardoso, S.V.; Sabino-Silva, R. Differential Molecular Signature of Human Saliva Using ATR-FTIR Spectroscopy for Chronic Kidney Disease Diagnosis. *Braz. Dent. J.* **2019**, *30*, 437–445. [CrossRef]
42. Rodrigues, L.M.; Magrini, T.D.; Lima, C.F.; Scholz, J.; Martinho, H.D.; Almeida, J.D. Effect of smoking cessation in saliva compounds by FTIR spectroscopy. *Spectrochim. Acta Part A* **2017**, *174*, 124–129. [CrossRef]
43. Stuart, B.; Ando, D.J. *Modern Infrared Spectroscopy: Analytical Chemistry by Open Learning*; Wiley: Greenwich, UK, 1996.
44. Naseer, K.; Ali, S.; Qazi, J. ATR-FTIR spectroscopy as the future of diagnostics: A systematic review of the approach using bio-fluids. *Appl. Spectrosc. Rev.* **2020**, *56*, 85–97. [CrossRef]
45. Derruau, S.; Gobinet, C.; Mateu, A.; Untereiner, V.; Lorimier, S.; Piot, O. Shedding light on confounding factors likely to affect salivary infrared biosignatures. *Anal. Bioanal. Chem.* **2019**, *411*, 2283–2290. [CrossRef] [PubMed]
46. Bel'skaya, L.V.; Sarf, E.A.; Solomatin, D.V. Age and Gender Characteristics of the Infrared Spectra of Normal Human Saliva. *Appl. Spectrosc.* **2020**, *74*, 536–543. [CrossRef] [PubMed]
47. Zijlstra, N.; Mars, M.; de Wijk, R.A.; Westerterp-Plantenga, M.S.; Holst, J.J.; de Graaf, C. Effect of viscosity on appetite and gastro-intestinal hormones. *Physiol. Behav.* **2009**, *97*, 68–75. [CrossRef] [PubMed]

MDPI
St. Alban-Anlage 66
4052 Basel
Switzerland
Tel. +41 61 683 77 34
Fax +41 61 302 89 18
www.mdpi.com

Foods Editorial Office
E-mail: foods@mdpi.com
www.mdpi.com/journal/foods

www.ingramcontent.com/pod-product-compliance
Lightning Source LLC
LaVergne TN
LVHW070658170726
843515LV00004B/440

* 9 7 8 3 0 3 6 5 6 6 6 8 9 *